Science & Islam

EHSAN MASOOD

Science & Islam

A HISTORY

ICON

Published in the UK in 2009 by
Icon Books Ltd, Omnibus Business Centre,
39–41 North Road, London N7 9DP
email: info@iconbooks.co.uk
www.iconbooks.co.uk

Sold in the UK, Europe, South Africa and Asia
by Faber & Faber Ltd, 3 Queen Square,
London WC1N 3AU
or their agents

Distributed in the UK, Europe, South Africa and Asia
by TBS Ltd, TBS Distribution Centre, Colchester Road,
Frating Green, Colchester CO7 7DW

This edition published in Australia in 2009
by Allen & Unwin Pty Ltd,
PO Box 8500, 83 Alexander Street,
Crows Nest, NSW 2065

ISBN 978-184831-081-0

Typesetting in 11.5pt Plantin by Marie Doherty

Printed and bound in the UK by
Clays Ltd, St Ives plc

Contents

Ehsan Masood is a science writer based in London. He writes for *Nature* and *Prospect* magazines and teaches international science policy at Imperial College London. He is a regular panellist on *Home Planet* on BBC Radio 4 and also presented 'Islam and Science', a three-part series for Radio 4 on science in today's Islamic world that was broadcast in 2009.

For my parents, Shamsa and Hassan Masood

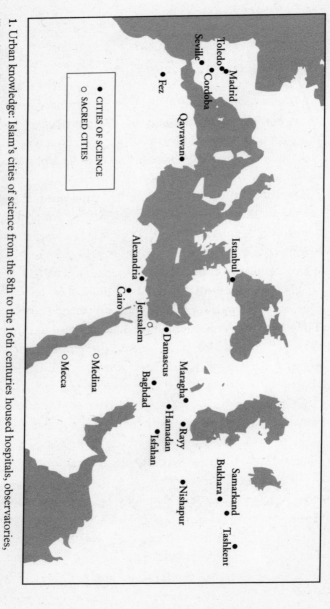

1. Urban knowledge: Islam's cities of science from the 8th to the 16th centuries housed hospitals, observatories, libraries, colleges and schools for translation, as well as much individual research.

A note on language

A book on science during the Islamic empires presents some interesting challenges for the science writer of today writing in English, and at a time when a good deal of sensitivity surrounds the use of words and phrases on all things Islamic, or Muslim.

Questions to do with God and religion are not mainstream to the process of how science is done, nor its many and varied products. Because of this, science writing (at least in English) has yet to develop a comprehensive vocabulary on the topic of science and religion.

The publishers of *Science and Islam: a History*, however, couldn't wait for a dictionary on science and belief. They needed a consistent short-hand phrase to describe the science that took place during the empires that followed the birth of Islam. And several candidates were shortlisted for the job.

One option was to use 'Muslim science', except that not all of the scientists mentioned in the following pages were Muslims. Another option was to go for 'Arab science', except that many practitioners were not from the Arab world, even if they were Arabic-speaking.

As so often happens with dilemmas in cultural relations, the best solution was to look for a compromise. The phrase that this book employs to describe science in Islamic times is 'Islamic science'. It isn't perfect by any means, but it comes closest to the mark.

An explanation of why Islamic science was chosen is needed, because to many readers Islamic science will be as nonsensical as Jewish science, Christian science, or Hindu science. Science is a universal tool for knowing about the world we live in: the individual beliefs of scientists have no bearing on the nature of what it is that they are investigating. One of the best examples of this is the 1979 Nobel prize in physics: this was shared between Muhammad Abdus Salam, a devout believer, and Steven Weinberg, a devout atheist.

To other readers, however, if something is 'Islamic' this means it is related to the practice of faith. To this group of readers, therefore, Islamic science might mean a science that is influenced by Islamic values, much as, say, Islamic banking is used to describe financial systems that are governed according to Islamic guidelines; or in the same way that an Islamic school is an institution that educates children according to Islamic values.

Just to be clear, Islamic science in the context of this book also includes a science that has been shaped by the needs of religion.

The second challenge relates to the word 'science' itself and what it means in languages such as Arabic, Persian and Urdu. The word 'science' in its modern context means the systematic study of the natural world, using observation, experimentation, measurement and verification. It comes from the Latin word (from around the 14th century) *scientia*, which means 'to know'.

Arabic manuscripts from Islamic times did not have a word for 'science' as we know it today. Instead, they had a word similar in meaning to *scientia*, which is *ilm* (plural, *uloom*). *Ilm* means 'knowledge': this could be knowledge of the natural world, as well as knowledge of religion and other things.

Scientists in the Ottoman empire came closest to realising that *ilm* is not the same as the scientific method. They introduced a new word, *fen* (plural, *funoon*), which means 'tools' or 'techniques'. For example, a university of science would be written in Turkish Arabic as *darul funoon*, or a home for the techniques of science.

The new Ottoman convention, however, did not catch on. Turkish Arabic is all but extinct and Modern Arabic has retained the original dual usage for the word *ilm*. So, while the Arabic edition of *Scientific American* magazine is called *Majalla Uloom* (magazine of knowledge), at the same time, *darul uloom* (a home for knowledge) is used to describe religious seminaries all over the world.

Those who continue to use *ilm* to mean both scientific and religious knowledge argue that it represents an idea (common to Islamic cultures) that science and faith are two sides of the same coin: that they are equally valid forms of knowledge, and with similar – if not equal – claims to seek the most truthful answers to questions.

Others would disagree. *Ilm* may well be an accurate description for religious knowledge; however, there is a case to be made for finding a word that can distinguish between scientific and religious knowledge.

In languages such as Arabic and Urdu, 'acquiring *ilm*' is a phrase that is commonly used in textbooks, in print and in the broadcast media. Knowledge of religion can of course be 'acquired' or memorised, as can much scientific knowledge. But science has an important added dimension: it is also about experimenting, innovating, building, refuting and pushing at the boundaries of what we know.

Prologue

Picture, if you will, images from the 1969 moon landings: those grainy black-and-white photos or slow-motion TV shots of rockets and astronauts in space, and awe-inspired spectators watching from below. Or recall the television footage from 2000 when the human genome had been sequenced, with the news announced jointly by US President Bill Clinton and Britain's Prime Minister Tony Blair.

What do these and so many more pictures of modern scientific discovery tell us? One message is very clear: that science is more than just 'science'. It is the result of the vision of those who govern us about where they want to take their societies in the future. The moon landings told anyone watching that here was an empire at the top of its game. Having established its domain on earth, the most technologically-advanced society of its age was ready to claim the heavens – or at the very least, a small part of it.

More than 1,000 years ago, another empire, that created by the coming of Islam, was at the top of its form.

This empire was in fact a network of what are called caliphates, united by a belief in God and in the teachings of the Prophet Muhammad. Its rulers and citizens spanned from Indonesia in the east to Spain in the west, and the last of the caliphates ended only in the last century, in 1923 with the fall of the Ottoman empire.

Science and Islam describes the scientific revolution that took place during the empires created by Islam, between the 8th and the 16th centuries. It is a story about the discoveries and inventions of a sophisticated culture and civilisation; the political and religious conditions surrounding it; and an extraordinary cast of characters – scientists, engineers and their patrons – who helped to make it all happen.

It describes an age when religion and science had a much closer relationship. Perhaps paradoxically, it was the needs of religion that in some ways helped to advance new knowledge. One example can be seen in efforts to develop quality standards in religious scholarship. After Muhammad's death in 632, scholars of religion wanted to find a way of checking and verifying the many records of his sayings. This led to a kind of early peer-review system, which later scholars of religion had to train themselves in. A century later, when scientific fields began to develop, it was theologians who encouraged the first scientists to adopt similar standards for authenticating their scientific work.

In its heyday, scientists and engineers from the Islamic world made groundbreaking discoveries and inventions, and we can see traces of their contributions today in our everyday lives. Moreover, many of the leaders of Islam's empires saw the relationship between science and society as would politicians in the modern age. They believed that the power of the mind could take us to places where no human had ventured in the past; they wanted the latest knowledge in order to help govern their territories and eliminate their enemies; and they wanted to shape societies in which people made decisions based on evidence and in which science, technology and rational thinking were important.

But then is not the same as now. Today's scientific endeavour is on a scale that is unprecedented in human history. In the United States, more is spent on healthcare research than the total spending of some of the world's poorest nations. And on the question of belief, today's scientists tend to keep their faith strictly private. Indeed, in the Western world at least, organised religion and individual religious faith are regarded by perhaps the majority of scientists as impediments to research, discovery and invention.

Today, knowledge is a highly specialised business: it is almost unheard of for a leading physicist to make ground-breaking discoveries in biology, or a chemist to push the envelope in philosophy. But many of the names that leap out of the pages that follow were polymaths who worked to the highest standards of the day.

In the pages that follow, you will meet many great thinkers such as ibn-Sina, a Persian-speaking scientist from the 10th century who also made important contributions to the study of the nature and philosophy of religious belief. He also found time to invent an early example of the micrometer, and his book *The Canon of Medicine* was taught to trainee doctors in universities in France and Italy from the 12th to the 16th century. Or there is Hassan ibn al-Haitham, an experimental physicist from the 11th century who helped to modernise our understanding of vision and who is credited with describing an early imaging device (a *camera obscura*), as well as writing about and researching the motion of planets.

You will also meet some of their patrons. Caliphs and governors such as al-Mamun of the Sunni Abbasid dynasty, who ruled from 813 to 833 from Baghdad, and al-Hakim of the Shia Fatimid dynasty who ruled from Cairo from 996 to 1021. These and many others employed personal scientific advisors, paid for libraries and observatories, and even personally took part in scientific experiments.

And you will meet some of the critics of the new science. These were men such as the theologian Abu Hamid al-Ghazali, who wrote a famous polemic, *The Incoherence of the Philosophers*, against the claims of scientists to be able to explain everything. And you will meet those scholars who suffered greatly for the right to criticise science and rationalism, men such as Ahmad ibn-Hanbal, who was tortured for refusing to accept that science should become the official religion of the Islamic state.

Turn the page and enter a brave, new and undiscovered world.

Ehsan Masood

1

The Dark Age Myth

If there is much misunderstanding in the West about the nature of Islam, there is also much ignorance about the debt our own culture and civilisation owe to the Islamic world. It is a failure which stems, I think, from the strait-jacket of history which we have inherited.

HRH Prince Charles in a speech at Oxford University,
27 October 1993

In 410 CE, Alaric, the Germanic king of the Visigoths, swept into Rome and sacked the great city in a three-day rampage. Sixty-six years later, Romulus Augustus, the last Roman emperor of the West, was deposed, and the regalia of empire was rudely despatched to Constantinople. With that, the lights went out on civilisation, and the Western world was plunged into an age of darkness – a night in which there was no scholarship, literacy or even civilised life. Only 1,000 years later did the world finally rediscover classical learning and bring the world's night of darkness to an end with the bright new dawn of the Renaissance. Or so the story goes.

This is the myth of the Dark Ages, the idea that history and progress pretty much stopped for a millennium after the fall of Rome. The trouble is that the myth is just that, a myth. But it has been a myth so potent that it has thoroughly distorted our understanding of how civilisations emerge and how science and learning progress. Advances in our understanding of the natural world happen when scientists absorb the latest knowledge in fields such as physics or biology, and then modify or improve it. They work rather like runners in a relay race, passing the baton of learning from one scientist to the next. Modern science, regarded as a hallmark of modern Western civilisation, achieved its place through the passing of many successive batons, which were handed to the scientists of Europe from those of the world's non-Western cultures. These included those who lived in the cultures of Islam over a period of some 800 years from the 8th to the 16th centuries.

The fact that we know little of this is what Michael Hamilton Morgan of the New Foundation for Peace speaks of as 'lost history'. The historian Jack Goody goes further and calls it 'the theft of history'. It is as if the memory of an entire civilisation and its contribution to the sum of knowledge has been virtually wiped from human consciousness. Not simply in the West but in the Islamic world too, the achievements of Islamic scientists were, until recently, largely forgotten or at least neglected, except by a few diligent specialists such as Harvard University's Abelhamid Sabra, David King, Jamil Ragep and George Saliba.

In mainstream science education in Britain – until very recently – the history of scientific progress has tended to leapfrog from the classical era of Euclid, Aristotle and Archimedes straight to the birth of the Age of Science in 16th- and 17th-century Europe, with only a cursory mention, if any, of the great swathe of Islamic science in between. In some versions of history, the 'dark age' only really ends, and the progress of science only really begins, with the famous conflict in the early 17th century in which Galileo confronts the Catholic Church with the assertion that the earth moves around the sun. As the world eventually acknowledges that Galileo is right, this is presented as the world-changing triumph of the light of reason over superstition. Thereafter, from the 17th century onwards, Western Europe's scientists are set free to unlock the world's secrets – William Harvey discovers blood circulation, Isaac Newton launches the study of physics, Robert Boyle pioneers the study of chemistry, Michael Faraday, electricity, and so on. And so we move forward into the Age of Reason and the dramatic progress of modern science.

Filling the gap

In reality, though, scientific inquiry did not simply stop with the fall of Rome, only to get going again in the 17th century. In fact, as this book will show, recent research is beginning to reveal just how thoroughly the 800-year gap was filled by a wealth of scientific exploration in medieval

Islam, and how it fed directly into the first stirrings of Western science.

The Cairo-based physician ibn al-Nafis, for example, discovered pulmonary circulation, the circulation of blood through the lungs, in the 13th century. Andalusian engineer Abbas ibn-Firnas worked out theories of flight, and is believed to have carried out a successful practical experiment six centuries before Leonardo drew his famous ornithopters. And in Kufa in Iraq, Jabir ibn-Hayyan (translated by Latin scholars as Geber) was among those laying the foundations of chemistry around 900 years before Boyle.

Moreover, some researchers are now showing that some of the great pioneers of modern science were building directly on the work of scientists from Islamic times. George Saliba of Columbia University, for instance, demonstrates in his book *Islamic Science and the Making of the European Renaissance* how the Polish astronomer Nicolaus Copernicus drew on the work of Islamic astronomers for the groundwork to his breakthrough claim in 1514 that the earth moved round the sun.

Historians of mathematics have also shown how algebra, a branch of maths that allows scientists to work out unknown quantities, was developed in 9th-century Baghdad by Musa al-Khwarizmi, building on work that he had discovered from mathematicians in India. Historians think that al-Khwarizmi would have had access to manuscripts through Islam's first encounter with India, which happened a century earlier. Modern science depends, too, on the solutions to complex quadratic equations devised by the poet and scientist Omar Khayyam. And much of our understanding of optics and light is built on the pioneering work of Hassan ibn al-Haitham (translated in Latin as Alhazen) in 11th-century Cairo.

Inventing the future?

The Islamic middle ages also left a strong legacy in the applied sciences. The nature of Islam, and the energy of a new empire, meant that there were many inventive and practical minds at work. According to Salim al-Hassani of the University of Manchester, some modern labour-saving devices such as the drinks dispenser could have an Islamic influence. Professor al-Hassani has recently introduced the world to some of the engineering achievements of al-Jazari, a 13th-century Turkish engineer, which include the crank, the camshaft and the reciprocating piston – all essential components of the modern car engine and much more besides. Meanwhile, a remarkable trio of irreverent but brilliant showman brothers, called Banu Musa, entertained 9th-century Baghdad with such ingenious trick machines and automatons that they would astonish even today.

If all these examples were fleeting moments of brilliance, they would be fascinating enough. But as many more teachers and historians are realising, they are much more than that. Names such as al-Khwarizmi and ibn al-Haitham are as integral to the history of science and technology as are Newton and Archimedes, James

Watt and Henry Ford, but the Arabic-sounding names somehow became lost in the myth of the Dark Ages. The reasons for this are the subject of an intense debate, which is as much about the relationship between the West and Islam as it is about the history of science and technology.

Lost in the dark

The idea that the Renaissance world was emerging from a period of darkness can be traced back to at least the 1330s, when the Italian historian Petrarch wrote of how it was that the world finally saw the light. 'Amidst the errors,' he said, 'there shone forth men of genius, no less keen were their eyes, although they were surrounded by darkness and dense gloom.' It may be that Petrarch was simply trying to link the emergence of Italian culture in his own time with its former heyday in Ancient Rome. But it is through men such as Petrarch that the notion of the dark ages was sustained, as Europe progressed towards the Enlightenment years of the 18th century and beyond. Perhaps tellingly, it reached its apogee, and acquired capital letters, when nations such as Britain, the Netherlands and Portugal introduced both Christianity and colonial rule into the continents of Africa and Asia. By this time, the Dark Ages had come to be seen as a time of decline into brute ignorance, full of 'rubbish', as Gibbon had earlier sneered in his *Decline and Fall*.

Perhaps it is no coincidence that this negative picture of the Dark Ages finally began to crumble along with the colonial empires. Many Western historians are now generally embarrassed by the distortion of history implied by the idea, and if they talk about dark ages, it tends only to be in a less pejorative sense, about periods that remain little known because of the dearth of written evidence. It is hard for them to see how an age that produced the *Book of Kells*, the scholarship of Alcuin and Bede, and countless great churches and monasteries could ever be thought of as an age of brute ignorance. More significantly, though, a tide of recent archaeological and textual research is now painting a much richer, fuller picture of life in Western Europe in the centuries after the fall of Rome, and even the idea that this is an unknowable period for the West is evaporating.

But of course the most distorting effect of the Dark Ages myth was the way it seemed to sideline, in the popular imagination at least, the history of the world beyond Western Europe – and virtually ignore the fact that learning had simply shifted eastwards, not completely flickered out. First of all, the Dark Ages myth seemed to turn a blind eye to the fact that the Roman empire did not actually end with the fall of Rome, but moved its centre to Byzantium. As the work of the historian Judith Herrin of King's College London shows so well, we are just beginning to wake up to the fact that cultural life – a rich cultural life – existed in Byzantium for the entire duration of the Dark Ages. And if Christian Byzantium was left in the shadows, equally telling has been the neglect of the achievements of early Islam.

The Dark Ages myth proved so powerful that even in some academic circles the best that could be said of Islamic scholarship was that it saved the great classical texts so that Europe could rediscover them in the Renaissance – as if retrieving them from a hole where they had been squirrelled away while the thousand-year storm blew past. With the old treasures retrieved, it was surmised, Islam was no longer needed and it was up to Europe alone to take knowledge forward.

Distorted imaginations

There are two main ways in which the Dark Ages myth has distorted the truth about the Islamic contribution to knowledge, culture and, in particular, science.

The first is the idea that the scholars of Islam acted as little more than custodians to the great classical works of scholarship, and added little of substance to the progress of human knowledge. Just how wrong that view could be will become clear later on. But it has led to much of the attention around the scholarship of the Islamic middle ages being focused on the 'Translation Project', the extraordinary movement to translate many of the great works of ancient Greece into Arabic, during Islam's so-called Golden Age under the Abbasid caliphs in the 9th century. This was indeed a phenomenal achievement, and it did ensure that the best of classical learning was not lost. But it seems likely that it is just one part of the sustained Arabic scholarship that began before the Golden Age of the Abbasids and endured for many centuries after, spreading well beyond Abbasid Baghdad, into Cairo and Cordoba, Persia and Uzbekistan.

There is what some scholars call a 'classical narrative' about Islamic science that has been put forward by orientalists in the past. This tells us that Muslim intellectual life shone for a few centuries under the Abbasids and their immediate successors. The Abbasids, led by the Caliph al-Mamun, were on the side of a progressive, rationalist approach to Islam, which enabled Muslims to take on board Greek learning in translation. But the growing influence of a conservative tendency which gave more weight to literalism in revelation and less to human logic gradually stifled scholarship. The turning point was a famous polemic against intellectuals in the 12th century by the theologian Abu Hamid al-Ghazali called *The Incoherence of the Philosophers*. When Baghdad and many other Islamic cities were destroyed by the horrific Mongol raids in the following century, Islam turned inwards and intellectual life declined – at just the time when the Europeans were able to go forward with the essentially Greek body of knowledge passed on by the Arabic scholars of the Golden Age.

The problems with this classical narrative are gradually being exposed, however. The philosophical standoff between the so-called rationalists and literalists was far more nuanced than it suggests, and the idea that Islamic science came to an end after al-Ghazali, or even the Mongol raids, is now known to be wrong. Some of the greatest minds of the Islamic era, such as al-Jazari, ibn al-Nafis and the astronomer Nasir al-Din al-Tusi, carried on the tradition well beyond the first stirrings of the European Renaissance.

An Islam *of* the West

The second distortion created by the Dark Ages myth is the notion that there was little or no positive contact between the West and Islam, and little real exchange of ideas, except for the eventual passing on of classical texts prior to the Renaissance.

There is no doubt that the history of the Crusades and the barriers of misunderstanding between the West and Islamic countries today help to reinforce the impression that beneficial contact between Islam and the West was minimal. It is highly likely, too, that many scholars in the Renaissance later played down or even disguised their connection to the Middle East for both political and religious reasons. The notion of the Dark Ages has reinforced the impression of separation. How could there be any contact between Islamic civilisations and a Europe lost in barbarian darkness?

Much new scholarship and archaeological research, however, is challenging this assumption. It now seems likely that there was considerable contact between Islam and the West even as early as the 7th century. In some ways, it is a mistake to talk about the Islamic 'world' and the Western 'world', as people often do; what is more accurate is to say that they are simply different parts of the same world.

The Arabs of Europe

For a start, Arabic-speaking merchants seem to have been trading throughout Western Europe at this time, providing wealthy people with luxuries such as sugar, carpets and silks. Gold dinar coins inscribed in Arabic have been found across Europe dating from the 8th to the 10th centuries. One of the most remarkable finds in England is a gold coin from the time of the Mercian King Offa, of Offa's Dyke fame, around 773–96 CE. This coin, now in the British Museum in London, is like the dinars minted for the first Abbasid Caliph al-Mansur in Baghdad in 773–4, with one exception – in the middle of the Arabic words stating that 'there is no God but Allah alone', there is inscribed in Latin capitals the name OFFA REX. A few scholars believe strongly that this could be evidence that Offa had converted to Islam. Equally likely, however, is that Offa had the coins copied – Arabic inscription and all – for the purpose of buying goods from the merchants of the Islamic world. An early example, if you will, of the idea of a trans-national single currency.

Across the English Channel at around the same time, the Frankish king Charlemagne was minting silver 'denarius' coins, also clearly modelled on Arabic dinars. He too was very much in the market for oriental luxuries. Indeed, Charlemagne was at this time exchanging gifts and letters with Baghdad's Harun al-Rashid, the caliph made famous in *The Thousand and One Arabian Nights* tales. In 801 CE, he sent Charlemagne an elephant called Abul Abbas, which is believed to have caused a sensation in the streets of Aix-la-Chapelle. The caliph also sent the king a carved ivory horn, a tray, a gold pitcher, a chess set, a tent, brass candlesticks and a water clock that astonished everyone who saw it and heard it striking the hour!

In addition, research from scholars such as Nabil Matar of the University of London shows that there was extensive and continuous contact between Islam and Christian Europe throughout the early and late middle ages in a host of different ways. Besides the merchants and entertainers who plied their trade across Europe, there was exchange of ideas and goods at every level in the places where the worlds of Islam and Europe became one – in Spain, in Sicily and in southern France – not to mention via Byzantium.

Some of the ways in which Islamic science and technology fed into Europe will be explored later in this book. At the same time, there are strong parallels between many things Islamic and those long regarded as part of the Western way of life. How they came to be will also be explored in these pages.

A shared Europe

It is already well known that coffee came from the East. According to one theory it was discovered after goatherds in Yemen, or perhaps Ethiopia (depending on which version of the story you read), noticed how frisky their charges became after eating certain berries. You might even know that the sugar that sweetens coffee originated here too. Indeed, there are many more everyday pleasures that are to be found in early Islam, but whose history is not so well known.

Take gardens as a place of relaxation rather than just a place for growing vegetables or herbs, for instance. They came to us from Persia. 'Early Muslims everywhere made earthly gardens that gave glimpses of the heavenly garden to come', says the historian A.M. Watson in his book *Agricultural Innovation in the Early Islamic World*. 'Long indeed would be the list of early Islamic cities that could boast huge expanses of gardens.' Islamic-era Toledo boasted Europe's first large botanical gardens in the 11th century. Many of the traditional flowers that grace the English garden also existed in the Islamic world – tulips, carnations, irises, and of course that quintessentially English flower, the rose. So too did many garden features, such as fountains and pergolas, conservatories and bandstands, not to mention mazes and sunshades.

Move indoors and you might walk across the carpet for a gentle game of chess. Both of these were also in use in the early Islamic world. Islamic carpets were imported as essential luxuries for centuries, long before the 18th-century Industrial Revolution meant that they could be made more cheaply in Europe. Chess, developed and played in India, came to Europe around the 9th century via Persia and Arabic-speaking Spain, and via the Viking trade routes from central Asia. The word 'checkmate' is similar to the Persian *shahmat*, meaning 'the king is defeated'. After your game, you might drink an aperitif from a glass – distillation and drinking-glasses are both innovations developed in Islam.

Even many deeper aspects of Western faith and culture are shared by those of the ancient Islamic world. The arches of some cathedrals, those pinnacles of Christian architecture, are shared by many mosques. And stained-glass windows were also used in Islamic times, as was the music notation: 'do, re, mi, fa, sol, la, ti, do'. Many of our basic institutions, too, can also be found in the world of medieval Islam, including public hospitals and libraries. The medicine of Hussain ibn-Sina (Avicenna) was Europe's default medical system up until the discovery of germ theory.

The cultures of Islam nurtured – and continue to have – a deep and rich tradition of love songs, poetry and romantic literature, some of which would undoubtedly have crossed over and synthesised with similar literary traditions in Europe. These traditions include the idea of doomed love – an early example of which is the 7th-century story of *Layla and Majnoon* and its countless variants, including of course *Romeo and Juliet*.

All of this might seem to have nothing directly to do with science, but the connection is important. Once you begin to appreciate something of the scope of the many links between Islamic and European cultures, it seems almost perverse to imagine that Islamic science and technology had no real effect on Western learning, and vice versa.

Part I

The Islamic Quest

2

The Coming of the Prophet

How could naked men, riding without armour or shield have been able to win ... and bring low the proud Persians?

Christian monk John Bar Penkayë, Turkey, 680 CE

The speed of the spread of Islam seems as astonishing today as it must have done when it happened in the 7th century. Alexander the Great and the Mongols also conquered vast areas very quickly. Yet both these conquests were short-lived and had little, if any, lasting effect. Islam seemed to change forever virtually every area it went to – from the west of China to the south of Spain, and including much of Asia, Africa and the Middle East.

Many of the lands that eventually came under Muslim rule would adopt Arabic and Persian as working languages, and Islam would become the new religion – permanently. Only Spain returned to Christianity after Muslims and Jews were driven out in the 15th century. However, the Arabic language has fared less well, and both Arabic and Persian today are confined to the countries of the Middle

East and North Africa. The majority of the Islamic peoples use indigenous languages, as well as European languages such as English and French.

Even more remarkable than the speed and permanence of Islam was the fact that it was achieved without a professional army: instead, Islam spread far and wide through the efforts of untrained men, often mystics, on camel and on foot. It's no wonder that they believed, as did many who witnessed it, that they were driven by a divine wind. The energy and vision which powered Islam's spread may well be the same motor that drove learning and science in the Islamic world, so it's worth looking at how it all began.

A lonely desert

In the early 7th century, the Middle East was dominated by two giant empires, the Sassanid empire of mostly Zoroastrian Persia in the east and the Christian Byzantine empire to the west. Both sprawled over vast areas – the Persian empire stretched far across Central Asia to the Himalayas, and the Byzantine empire wrapped almost right around the Mediterranean – and they had seen few challenges to their dominance for centuries. But rivalry, and perhaps a series of plague epidemics, had sapped their strength. Byzantium had only just recovered, after a long and bloody struggle, the former territories in the Levant, the eastern Mediterranean, that it had lost to the Sassanians. And maybe they were too preoccupied with each other to pay much attention to their southern fringes.

Here, sandwiched between them, lay the lands of the Arabic-speaking peoples. Arabia was then, as it is now, vast yet sparsely populated, much of it desert where rainfall is almost as low as anywhere in the world. The few precious patches of green are almost lost amid swathes of scorched sand and gravel, and barren, buff-and-grey plateaus, cliffs and gorges. By day, the sun blazes down here relentlessly and twice every year, for a month or more, winds blow from the north, picking up hundreds of millions of grains of sand and dust. These storms billow and roar across the desert in a blinding, stinging cloud, or spin off into vicious little whirling, twisting dust devils named *djinn*, after the magical spirit-beings of Arabian folklore, known in English as genies.

It is perhaps no wonder, then, that neither Persia nor Byzantium felt the need to conquer this challenging land. And so the empty spaces of Arabia were left largely to the peoples of the desert, described by Western scholars as the Bedouin. Some moved from oasis to oasis with their livestock, while the few permanently settled areas and towns were occupied by clans or tribes. Uninterested, the empires simply relied on intermediaries to keep these tribes in line.

Both the nomadic and the settled populations worshipped a range of lesser gods that they believed were subordinate to a supra-God, whom they called Allah, the High God. They had a network of leaders; and boys and young men were taught from an early age how to ride and were expected to become skilled swordsmen and archers. At the same time, a rich tradition of romantic oral poetry glorifying military heroism also emerged. This pre-Islamic warrior culture was later described by the Muslims as the *jahiliya*, or age of reckless ignorance.

Trouble in Mecca

As the Sassanian princes reclined on their carpets and cushions in Ctesiphon, the Persian capital, and the Byzantines sucked on the fruits of empire in Byzantium, they probably thought little of the effects of their taste for luxuries. Yet across the Persian Gulf, the trade in spices and precious metals and gems through the Arabian peninsula was having a hugely disruptive influence on Arabia. As merchants competed for business in trading towns, an uncomfortable social divide was opening up between rich and poor.

It was in one of these towns, Mecca, that a young merchant called Muhammad, born in 570, began to worry about the consequences of the pursuit of wealth for its own sake. Historians disagree over whether or not Mecca was a spice town, although it was certainly a thriving trading centre and Muhammad's tribe, the Quraysh, had something of a monopoly over the goods caravans that moved between Syria in the north and Yemen to the south. But Mecca had another, more significant claim to fame.

The stone and the well

Although not far from the Red Sea port of Jeddah, Mecca was not what would you call a verdant oasis, located as it was in a desolate valley between mountains. But most people didn't come to Mecca for the scenery; they came as pilgrims to visit the Ka'bah. The Ka'bah is now the most sacred place on earth for Muslims, located right at the heart of the Sacred Mosque, and it is towards this that they turn when they say their daily prayers. But it was a sacred spot long before the coming of Muhammad, and is believed by Muslims to have been built by Abraham as a place to worship a single God. Outside the cubic shrine are the remains of a black meteorite called the Black Stone of Mecca.

Just a few metres from the Ka'bah is another sacred site, the ancient well of Zamzam. According to Islamic tradition, the well was miraculously revealed to Abraham's second wife Hagar 4,000 years ago. The family paused here on a journey south while the ageing Abraham went back for his first wife Sarah. But the place where he left her was dry, and Hagar was soon desperate for water for her infant son Ishmael. As she scoured the area in a panic, Ishmael began to dig the ground with his foot. The result of this was that water gushed from the ground and Hagar had to build a dam of sand and stone to stop it flowing away. The name Zamzam comes from the phrase *zomë zomë*, meaning 'stop flowing'. When Abraham returned, it is said, he built the Ka'bah nearby.

The young reformer

So, Mecca was a place of pilgrimage long before the coming of Islam. The city is also mentioned in the Bible in Psalm 84, as the well of Baca. But by Muhammad's time, belief in a single God was replaced by belief in many, which was reflected in the fact that the Ka'bah contained representations of these many deities.

Muhammad is described in Islamic literature as being an idealistic, upright, but trusted young man born to the Quraysh elite. He is said to have hated the venality and superficiality of Mecca even before revelation. He and his friends would often do things to help the poor and the old: widows, orphans and slaves. A wealthy widow named Khadija, clearly struck by the honesty and integrity of this young merchant, employed him as manager of her business affairs, and later proposed marriage to him. The marriage was long and is said to have been a happy one. Yet he continued to be repulsed by the greed and poverty he saw in his home city, and as he approached middle age, he became increasingly upset by it.

Revelation

In 610 CE, when Muhammad was 40, he was sitting in the cave beyond the city limits on Mount Hira where he often retreated to meditate. It is here that Islamic tradition says he saw a vision that turned out to be the angel Gabriel. The angel said to Muhammad:

Read in the name of your Sustainer who created you.
Who created humans from congealed blood.
Read and your God is most bountiful.
Who taught through the pen.
Who taught humans what they did not know.

(Qur'an, verse 96)

The angel's words would later become an anthem for science and learning in the Islamic world.

More revelations came, and Muhammad began to preach to members of his family and then to others in Mecca. He called on them to accept a single, all-powerful God, to reject the need for intermediary gods, and also to do away with greed and to treat fellow citizens with justice and to give them dignity. Khadija immediately joined him and Muhammad quickly found other followers, drawn by his denunciation of greed and his vision of equality. He stressed from the start that he was not starting a new religion, but was simply reminding people to revert to the teachings of prophets who had come before him, and that Allah was the God of Abraham, Jesus and Moses.

Even so, the wealthy and politically powerful of Mecca found his ideas threatening, and in 622 Muhammad fled the city for Medina, an event called the *hijra*. Yet this proved to be just a temporary setback. Such was the appeal of his ideas that converts came quickly and in large numbers.

Warriors and martyrs

From the beginning, Islam's survival depended not just on the mouth-to-mouth spread of ideas. Islamic tradition says that Muhammad tried to avoid armed conflict, but that he and his fledgling group of followers in Medina were forced early on to fight for their survival against attacks from the Meccan elite. A series of skirmishes culminated in the battle of Badr outside Mecca, the first battle for Islam, in which a small Muslim army triumphed over a much larger Meccan army. For the Muslims, this victory was powerful proof that God was on their side.

The belief that God was helping them on to victory came to play a large part in success on the battlefield. Believing that they had a divine wind at their back gave the Islamic armies the extraordinary confidence that enabled them to sweep aside much larger and better-equipped forces. From the first accounts of the battle of Badr, Muslims who died fighting for the cause were considered martyrs, and as martyrs they would go straight to heaven. It is likely that some of the early Muslims were driven on, not just by the promise of the next world, but by the promise of rewards on earth too. Many of those who fought for Islam in the early years were among the poorest people in the region. The spoils of the Byzantine and Sassanian empires would undoubtedly raise their standards of living.

The Islamic state

Politics has always been important to Islam. From Muhammad's time, the first Muslims found themselves ranged against those who held political power, initially because they were not allowed to preach openly, and later to defend themselves against attacks from the Meccan establishment. A decade after the first revelation, however, Islam would itself become the source of power. Muhammad would establish rules for a small Islamic city state in Medina, and later generations of Muslims would seek to use myriad interpretations of Islam as the basis for exercising power across the new Islamic empires.

This political aspect of Islam would also help to drive the later championing of science. As in all empires, science and knowledge in the Islamic empires were all part of political power. Islam was both a faith and a political movement. Muhammad's aim was not just to provide individuals with a new way of life, but to create a new vision of society – a new state and community. This is one reason why it has been so especially hard for those Muslims who were later colonised by the nations of Europe to live under the authority of the new rulers – rulers who Muslims believed did not share their view of a just society.

After Muhammad

When Muhammad died in 632, Sunni Muslims believe he did so without formally anointing a successor.

Confusion about who should take over the leadership of the new movement caused many problems – problems so profound and lasting that they have led to many centuries of conflict and remain at the heart of many of the tensions that exist in the Islamic world today, including the divisions between Shia and Sunni Muslims.

There were at least three groups from which a new leader could have been chosen: he could have come from one of the faithful Companions who had accompanied Muhammad from the earliest days; it might have been someone from the Quraysh tribe; or Muhammad's own family – in particular, his cousin and son-in-law, Ali. Ali's supporters would later establish Islam's Shia branch, distinct from the Sunni majority. In the end, the role of Muhammad's first *khalifa* (successor, or caliph) was given to one of his closest Companions and one of the first Muslims, Abu Bakr. Neither of the groups that lost out was completely happy about this. The first four caliphs – Abu Bakr, Umar, Uthman and finally Ali himself – were later called the Rightly-Guided Caliphs, and each caliph was chosen anew rather than the caliphate being passed on to their children. However, Abu Bakr was the only one of them who did not meet with a violent end.

It was probably to stabilise that situation that a powerful Meccan family called the Umayyads, led by Muawiya, eventually took control after Ali was murdered in 661. The Umayyads moved the caliphate from Mecca to Damascus and with Muawiya established the first dynasty of caliphs, which was to last until 750 when they were ousted by the Abbasids, a family with connections to Ali.

Yet despite the tensions at the top, in less than three decades under the first four caliphs, the forces of Islam had conquered both the entire Persian empire of the Sassanians and much of the Byzantine empire. Under Abu Bakr, in just two years after Muhammad's death, they had consolidated their power over all Arabia, taken Iraq from the Persians, and won Damascus from the Byzantines.

Perhaps the greatest of the Islamic victories was at Yarmuk in what is now Jordan in 636, in which an army of some 30,000–40,000 routed a Byzantine force of well over 100,000. At one point during the battle some of the Muslim soldiers withdrew, it is said, only to be met by their women who drove them back into the fray with tent poles, singing:

> O you who run from a constant woman
> Who has both beauty and virtue;
> And leave her to the infidel,
> The hated and evil infidel,
> To possess, disgrace and ruin.

Without that kind of pressure behind them, perhaps it's no wonder that the Byzantines were routed. The battle was a disaster for them, and the Byzantine emperor Heraclius fled by ship from Yarmuk back to Byzantium, apparently blaming the defeat on divine vengeance for his marriage to his young niece Martina.

The Muslim victory at Yarmuk sent shockwaves around the world. Some time later, the monk St Anastasius noted from Mount Sinai that it was 'the first and fearful and incurable fall of the Roman army'. After Yarmuk, the Byzantine empire quickly lost much of its territory and was reduced to a rump in what is now Turkey and around Byzantium. In 638, the Muslims captured Jerusalem, a moment of great symbolic importance. By 640, most of Syria was under their control. Two years later Byzantine Egypt fell, and within a few decades so had all of Byzantine North Africa as far as the Atlantic. Meanwhile, just a year after Yarmuk, the Persians had been defeated in their heartland of Iran after the battle of Qadisiyyah, and soon after, Muslims had pushed right through Persia to the borders of Central Asia.

3

Building Islam

Even if you must go all the way to China, seek knowledge.
The Prophet Muhammad

By the time the Umayyad Caliph Muawiya came to power in 661, the task of the leader of Islam had changed dramatically. For the first few decades after the Prophet Muhammad had died, the new Muslims had been preoccupied with battles against their Persian and Byzantine neighbours. In later years, the caliph's priorities changed and he turned his attention to administering an empire.

He still needed an army to police the empire and defend its territory – but this would now be a professional army of paid soldiers from any religion or race, and not a voluntary army. The caliph also needed administrators to collect taxes and run local affairs, and he needed skilled men to maintain the empire's infrastructure. And the caliph himself needed to be at the most practical place to run an empire – the major city of Damascus, at the centre of the empire, not Mecca, a small town on its south-eastern fringes. Damascus became the capital and very quickly this ancient city, already well over 6,000 years old, was the bustling hub of the Islamic world.

The Muslims turned to the task of maintaining their new empire with the same energy that had driven them to creating it. The consensus among historians is that little real effort was made to convert those non-Muslims who found themselves citizens of the Islamic empire. As the experience of the Soviet Union and China amply demonstrates, lasting belief in a religion or an ideology can happen only through free will and cannot be forced. For a long while, less than a tenth of the Islamic world's population was Muslim. The other 90 per cent was a mixture of Christians, Jews, Zoroastrians and those from other faiths. Christians in the Islamic empire on the whole found that they were treated better by the Muslims than they had been by their former Byzantine rulers. Byzantium adhered to a form of Christianity which insisted on a single interpretation of the nature of Christ, and they persecuted dissidents such as the Nestorians, who gradually escaped eastwards from Syria through Persia, eventually reaching China (the first Christians there). Nestorian and other Syriac-speaking Christians were later to play a key role in the project to translate works of learning from Greek to Arabic.

Historians argue that the first Muslims were not keen on too many new converts partly because of money. The majority People of the Book – that is, Jews, Christians and others whose faiths were based on sacred texts – had to pay different taxes from Muslims, which would guarantee them the protection of the state and exempt them from military service. So encouraging a much larger number of Christians and Jews to become Muslim could have meant a loss of revenue for the rulers. However, Christians and Jews, by the standards of today, did experience discrimination. They were not allowed to stand for the highest office, and they were not entitled to free healthcare in some places.

The Umayyad mosque

In the meantime, such a multi-cultural and multi-religious atmosphere meant that many people were able to bring their energies to bear in the dynamic new empire. In few places is the uniquely Islamic mix of talents clearer than in the Umayyad mosque in Damascus. For the first 40 years in Damascus, Muslims simply shared a small church with the Christians. Then in 706, the Caliph al-Walid I bought the church from the Christians, had it demolished and built a mosque. This was the first great Muslim building. The ambition and organisational skill is clear in its scale.

When Caliph al-Walid launched the scheme, he announced: 'Inhabitants of Damascus, four things give you marked superiority over the rest of the world: your climate, your water, your fruits and your baths. To these I wanted to add a fifth: this mosque.' What he gave them with the mosque was one of the biggest buildings that had been built since Roman times, and one that still looks striking after almost 1,300 years. Tellingly, maybe, for the future development of science, the mosque absorbed the latest classical architecture and then moved it forward. Al-Walid even shipped in 200 of the best Byzantine craftsmen to provide him with some wonderful mosaics. But then these elements were transformed into something new and uniquely Islamic. The confident expression of an entirely new Islamic style of building and decoration, from the elaborate geometric swirls to the richly coloured walls, the domes and minarets, is unmistakable.

Progressing in faith and science

Absorbing the best of other civilisations and then modifying and innovating with new ideas is the hallmark of science. It is also one of the characteristics of Islam. Muhammad made it clear that Islam was not a new religion. Christianity and Judaism both came from the same root, and like Islam they looked back to Abraham as their starting point. The Qur'an was not the only word, simply the last. In the Islamic tradition the Prophet made a miraculous night-time trip to heaven from Jerusalem, called the *mi'raj*. While in heaven, Muhammad met all the prophets, including Jesus, and led them in prayer. Islam's calling-card was that it provided both a link to the past and a confident new future.

During the years of the Umayyad caliphates, it seems that the administrative and practical machinery of the empire was developing rapidly. Moreover, there was a large economic stimulus. As Muslims spread out across the Middle East and into Africa, for instance, tent-cities were established at places like Basra and Kufa in Iraq and Fustat in Egypt. All these cities eventually became permanent – Fustat developed into Cairo – but from the beginning they needed supplying with all kinds of goods, and the new armies, administrators and their families had the money to pay for them.

The farming revolution

In particular, they needed food, and the fertile lowlands of Iraq, and later the plains of Egypt and Andalusia, became the focus of what can only be called an agricultural revolution. The Arabic-speaking peoples had always been great travellers, and as the empire expanded, they brought ideas from Morocco to Mongolia for boosting food production in their new homes. All these ideas were eventually incorporated into farming manuals.

From Andalusia, for instance, the Muslims of Iraq discovered and then developed crop rotation. Previously, there had been just one harvest every year in winter. With crop rotation, they were able to obtain several harvests every year. But this innovation would not have been possible without another. The dry, hot weather that often afflicted the Iraqi lowlands would have made summer cropping impossible. So irrigation techniques were developed. Sugar, for instance, had to be watered every four to eight days in the summer, but astonishingly, the early farmers achieved it.

The introduction of the famous ancient water tunnels or *qanats* from Iran was just one of their successes with irrigation. Even more impressive were their water-raising techniques, and in particular their *norias* or water wheels. The first mention of *norias* comes from the time of the Umayyad caliphates in connection with a canal being dug near Basra. Although the famous *norias* of Hama in Syria date back only to the 14th century, they are probably typical of countless water-raising machines that were soon in use all over the Islamic world. Water-raising later became the focus of some of the key technological achievements of the Islamic era, such as those by al-Jazari, the Turkish engineer, as we shall see later.

New crops and new owners

Besides new techniques and water management systems, crops were taken from one part of the world and introduced elsewhere. Oranges and lemons, for instance, came from India to the Middle East around the end of the 9th century and soon spread across the Islamic world and into Spain. In the same way, the empire cultivated and then spread sugar, pomegranates, figs, olives, cotton and many other crops far and wide.

Many of these innovations needed another key development: this was the idea of property rights for small farmers. The Islamic empires were not feudal states, and individuals were allowed to own land so long as they paid taxes on it. This kept the cities well fed and also contributed to the exchequer. The new complexities of land ownership, calculating appropriate shares, working out tax bills and so on were probably also a key factor in the pressure from officials right up to the caliph to develop mathematical and computational systems to handle them. Many of the examples that the mathematician Musa al-Khwarizmi was soon to use to demonstrate his new technique of algebra come from the world of farming and landowning. And the need to get accurate information for planting and harvesting times may have had a similar effect on astronomy.

A new language

The administration of the Islamic territories was probably an equally important stimulus to scholarship and science. For the first few decades, the business of government was performed in the relevant national language, with the aid of interpreters. In the 690s, however, Caliph Abd al-Malik decreed that Arabic was to be used in all official documents. That meant that anyone wanting to work for the government – even conduct business with government officials – had to be able to write Arabic. The long-term impact of this simple measure was huge. Gradually, pretty much everyone from Andalusia to Afghanistan learned to speak a form of Arabic. For anyone who could write, Arabic became effectively a universal language right across the vast extent of the Muslim world. Just as English is the language of science today, so the spread of written Arabic allowed scholars from distant places and numerous different cultures to communicate their ideas easily and note them down for others to read. It's probable that this, as much as anything, helped to enable and sustain Islamic science over so many centuries.

Making a mint

At around the same time, Abd al-Malik introduced another far-reaching measure. Just as with language, the Muslims had made do with existing Byzantine coinage in the early decades. Apparently, one story goes, the Byzantine emperor was unhappy with a mention of the Prophet in some official documents, and countered by threatening to have coins inscribed with words that would offend Muslims. So, the story goes on, Abd al-Malik asked for advice from a famous prince and scholar called Khalid ibn-Yazid. Khalid's solution was simple – make your own coins. And from that time on, dinars, complete with an inscription praising God, became the currency for the new empire.

Historian George Saliba suggests that events such as this might have proved a stimulus for the start of the movement to translate scientific texts into Arabic and for scientific experiment in general. 'If this anecdote is taken together with Khalid's expressed interest in alchemy,' Saliba says, 'we can see why such books on alchemy may have come in very handy to someone who was interested in striking new mint of gold coins. Who but the alchemists would be better prepared to identify pure gold, from other metals? And who but the alchemists would be the experts who could judge alloys and the like?' Whatever the truth of Saliba's suggestion, Khalid is generally thought to be the first of the Islamic alchemists, and among the first to initiate the translation of scientific texts into Arabic.

The Umayyads and their discontents

Despite the gradual economic development under the Umayyads, however, all was not well on the political front. In the east of the empire, especially the far east where the ancient Persian empire stretched into Central Asia, there were discontents. Some were those in Persia, both Muslim and non-Muslim, who disliked the hold on power by Arabic-speaking peoples. Some were those who felt that the family of Muhammad had been deprived of the caliphate by the Umayyads. Others just felt aggrieved about the relentless flow of cash and resources westwards to Damascus. In fact, there were probably innumerable reasons for discontent.

Many could have exploited this discontent, but it was the Abbasids who did. The Abbasids were supporters of the side of the Prophet's family descended from his uncle, Abbas. They were mainly based in Kufa in Iraq, but they sent out agents and emissaries to build up support on the eastern fringes. Their message was simple: that if the Umayyads were ousted and the Family of Muhammad put in their place, the world would be a better place. Finally, in the summer of 747, the revolutionary black banners of the Abbasids were unfurled in Merv, the ancient oasis city in Khorasan in the middle of the Kara

Kum (Black Sand) desert. Led by Abu Muslim – possibly a pseudonym, since it means 'Father of Muslim' – the revolutionary army marched west, swelling in size as they scored victory after victory over the Umayyad armies, who had no particular allegiance to the caliph.

Once they reached Kufa, the Abbasid candidate Abu'l-Abbas declared himself caliph. From Kufa, the army drove on westwards, winning skirmish after skirmish, until finally they met the Umayyad Caliph Marwan and his forces at the River Zab near Mosul in northern Iraq in February 750. Marwan was routed, many of his fleeing troops being drowned in the Zab river, swollen by winter rains. Almost alone, Marwan was pursued across Syria and on south, until finally the revolutionaries caught up with him and killed him.

The age of the Abbasids had begun.

4

Baghdad's Splendour

To [Baghdad] they come from all countries far and near, and people from every side have preferred Baghdad to their homeland ... There is none more learned than [the Baghdadis'] scholars, better informed than their traditionalists, more cogent than their theologians ... more poetic than their poets, and more reckless than their rakes.

Ahmad al-Ya'qubi, writing of a visit to Baghdad in the time of Caliph al-Mamun, 9th century

With the coming of the Abbasid caliphate, the curtain rose on the spectacular setting for what some have called the Golden Age of Islamic science – the city of Baghdad. Keen to start afresh, the second Abbasid Caliph al-Mansur abandoned Damascus and set about creating an entirely new capital nearer the heartland of their support in the east, right in the middle of the newly productive farmland beside the Tigris and Euphrates. And what a city they built.

Virtually nothing of the Abbasid city of Baghdad survives today to show what it was really like, but there is no shortage of Arabic written sources to testify to the city's glamour in Abbasid times. Within just a few decades of its founding in 762, it had grown into one of the world's greatest cities, not simply in terms of size – estimates suggest that it had a population of up to a million, at a time when few cities outside China had anything more than a few tens of thousands – but in terms of the bustling, energetic, cosmopolitan mix of people who came here from far and wide to live and work.

Baghdad is of course the city fabled in *The Thousand and One Arabian Nights*, where Scheherezade spun the tales to enchant her prince, the third Abbasid Caliph Harun al-Rashid – a city of fountains and courtyards, of carpeted and cushioned rooms where silk-clad girls would dance and poets would yearn for them, of clandestine meetings and light-fingered thieves in the night. There is no way of knowing now how much of this is true.

The Round City

At the centre of Baghdad stood a perfectly round city surrounded by a high wall. At each quadrant of the circle was a giant gate, through which roads led to the four corners of the empire – to Khurasan, Basra, Kufa and Syria. Much of the Round City was an empty park with the royal palace and mosque at the centre. The vast, sprawling mass of the city, with its narrow *suqs* and winding streets and its flat-roofed courtyard houses large and small, lay outside. The circular layout mirrored the classic Persian city of Fairouzabad, and the domes and arches of the palace may have been inspired by the great palace of the Persian shahs just a few kilometres away at Ctesiphon, the grand arches of which still stand.[1]

The tone of the city, too, in the early years was set by an established aristocratic family from Balkh in Afghanistan called the Barmakids. The Barmakids were among the richest families in Baghdad by far, and by all accounts were shrewd political operators. Three generations of the family became advisors to the Abbasid caliphs and effectively ran the empire, controlling not only the empire's finances but also who got to see the caliph and who didn't. One advisor is believed to have held the keys to the caliph's harem. Eventually, though no one knows quite why, it seems that the Barmakids overstepped the mark, and in 803 Harun killed the last Barmakid advisor Ja'far – once his companion in many youthful escapades.

Baghdad style

While they were in power, the Barmakids set the tone of patronage of the arts and science followed by many other of the city's wealthy, and of arranging *majalis* (salons) in which courtiers and scholars got together to

---

[1] Apparently, al-Mansur was not only inspired by the architecture of Ctesiphon but wanted to use its very bricks. But Khalid ibn-Barmak, his advisor, suggested that the continued existence of the Sassanian palace as a ruin was the perfect reminder of the superiority of Islam. And so the palace ruin was spared to survive to this day.

debate religious and philosophical ideas with remarkable openness. Not just Muslim scholars of all persuasions, but Christians, Jews and Zoroastrians were welcomed at these salons – and the only criterion for entry seemed to be whether you could argue your case well. All the same, it wasn't entirely free from rivalries and prejudices against outsiders.

There is a story about the young Hunayn ibn-Ishaq (Johannitius in Latin), later perhaps the most famous of translators and a medical doctor. Hunayn came from the country village of Hira in Iraq, but like so many young men of the time went to Baghdad to make his name in medicine. There he went to the *majalis* of the court physicians to four Abbasid caliphs. In his eagerness, young Hunayn kept on questioning what the learned physicians at the salon were saying. Eventually, one of the old medical elite from Persia became so irritated by the young upstart that he threw him out as a country bumpkin. 'Go change money in the streets!' he is reported to have said. Years later, when Hunayn had proved himself, the physician apologised.

Greek topics

It is likely, though, that there was considerable prestige attached to success in the salon and in displays of learning and erudition. Perhaps significantly, one of the first of Aristotle's works to be translated into Arabic was his *Topics*, in which he gives advice on how to argue a case. What better way to upstage your rivals in debate than

learning from, and quoting, the master? Maybe one reason why the wealthy elite were willing to spend fortunes
on getting translations of learned works into Arabic was
simple one-upmanship?

Amira Bennison of Cambridge University suggests a
grander purpose for translating Greek philosophy. She
thinks it may have been that, as a new empire and religion, Islam wanted the tools to develop its own theological and philosophical arguments to use against Christians
and Jews who already had well-developed dialectic traditions. Peter Adamson, on the other hand, suggests a
simpler nationalist motive – effectively, Muslims wanted
to eclipse the Byzantines by showing that they understood and appreciated their own Greek heritage better
than they did. Meanwhile, Yahya Michot of the Hartford
Seminary in Connecticut and Dmitri Gutas of Yale
University add yet another incentive for Abbasid caliphs
to pay for translations – astrology, which both gave them
the power of prediction and confirmed the legitimacy of
their revolution because it was written in the stars. Other
historians of the era suggest that the Persians were simply
reclaiming their heritage of knowledge after Alexander
the Great destroyed Persian Persepolis in 330 BCE.

The translation movement

Whatever the reasons, and they were probably many, the
extraordinary boom in translation that ran through the
Abbasid caliphate seemed to be in line with Muhammad's
injunction to 'seek knowledge everywhere, even if you

have to go to China'. And Baghdad, with its new-found wealth and the extraordinary cosmopolitan mix that the spread of Islam gave it, was the perfect venue. There was wealth to be made in the translation business, too. The famous Banu Musa brothers were apparently happy to pay translators 500 dinars a month – the equivalent of about £24,000 today – and the wealthy elite would have to pay much more to get their own personal translation of one of Aristotle's major works. It was a price they were willing to pay.

The translation movement began slowly in the caliphate of al-Mahdi (775–86) and Harun al-Rashid (786–809) but really got under way in the time of al-Mamun. Soon, ancient manuscripts were flooding into Baghdad (and also Basra), ready to meet the demand. Most were Greek, but they also came from Persia, from India and maybe even China too. Al-Mamun and members of the elite sent out missions to find manuscripts. One possibly apocryphal story tells of a mission to Byzantium, where only after some persistence the manuscript hunters were told that they might find Greek manuscripts locked away in an old church. When they finally gained entry, they found many of the great pieces of Greek scholarship in a fragile state, covered in dust and cobwebs or even rotting away with mildew. Some of the manuscripts for translation were recovered by official missions in this way. Others were, no doubt, brought to Baghdad by those eager to cash in on the booming market.

Many of the translators were not Arabs, nor even Muslims, but from faiths and languages brought in by the huge expansion of the Islamic realm. Many were native Greek speakers from the old Byzantine empire. Many, too, were Christian scholars who spoke and wrote Syriac, a form of Aramaic, the original language of the Bible and adopted by many eastern Christians between the 4th and 8th centuries. Many of the texts were actually translated first into Syriac and only then into Arabic in a two-step process. Indeed, some of the Greek texts, such as Aristotle's, already existed in Syriac. In the preface to one major translation of Aristotle into Arabic, the Syriac Christian translator tells how his Arabic was corrected by his boss, al-Kindi.[2]

The language factory

Greek authors were the primary targets, though texts from other languages – such as those from South Asia – were translated too. The range of texts that were translated was wide. On the whole, though, it was almost entirely scholarly texts, rather than works of literature. Top of the list were subjects that had a clear practical use – medical texts such as those of the famous Galen and Hippocrates, mathematical texts such as Euclid's

---

[2] Surprisingly, perhaps, Christians and Muslims found that they had quite similar outlooks when it came to translating the Greeks. Both Christians and Muslims believed in one god; the Greek philosophers were pagans who believed in many gods, or disregarded the gods altogether. So often Christian and Muslim translators would adjust the Greek text in similar ways to make it more palatable to their potential readers.

Elements, and astronomy texts like Ptolemy's *Almagest*.[3] (Just why astronomy was so important practically to Muslims will become clear in Chapter 9.) Philosophy was also a popular subject, especially the works of Aristotle and Plato.

What has surprised scholars in the modern age is the speed with which the Abbasid translators learned to translate highly technical texts accurately and fluently. This has led some to argue that the translators were already very familiar with the subject matter. They included translators of many nationalities, and may also have included many such as the Persians who had long been familiar with astronomy.

It was a real challenge for the translators to find words in Arabic to correspond to the range of technical terms in the original, yet they were inventive in their solutions and Arabic soon had its own sophisticated technical vocabulary. Peter Pormann of Warwick University cites an example of how Arabic terms came about. The disease alopecia, for example, got its Greek name because it resembled mange in foxes, which is *alopek* in Greek. So the Arabic translators called it in Arabic 'the fox disease'.

Writing in the 14th century, al-Safadi said that the translators had two basic approaches. One was literal, in which the text was translated word by word, trying to find the equivalent each time in Arabic. The second was

---

[3] 'Almagest' is the Latin transliteration of the Arabic name for Ptolemy's book, which translates as 'the greatest'.

an approach in which the translator instead tried to con-
vey the intention and meaning of the writer. Al-Safadi
points out how the literal translations were often incom-
prehensible, and that much better results were achieved
with the second approach pioneered by Hunayn, who
became one of the most famous of the translators.

Hunayn: medicine man

Hunayn was a Christian, and after his salon argument
with the court physician, he set off to Byzantium to learn
Greek and Syriac. When he returned to Baghdad a few
years later, still only seventeen, he was commissioned by
a court official to translate the works of Galen. It was
to Hunayn more than anyone that later ages owed the
survival of so much of the work of Galen, the first great
medical texts and the basis of most medical knowledge
for 1,000 years.

However, Hunayn was not content to merely translate.
He was a doctor, and where he saw limitations in Galen's
work, he improved on it. He made crucial additions to
Galen with his anatomy of the eye, and his drawings are
typical of the wonderfully clear scientific illustrations that
became a hallmark of the Islamic scientists' work, and of
the best science texts ever since. Hunayn also wrote a
brief summary of Galen's work in question-and-answer
form, which was one of the first Arabic scientific texts to
be translated into Latin in the 11th century and became
a key medical primer for many centuries.

Hunayn fell from grace later in life, but his translations made him rich and highly respected. One observer from the time describes his lifestyle:

He went to the bath every day after his ride and had water poured on him. He would then come out wrapped in a dressing gown and, after taking a cup of wine with a biscuit, lie down until he had stopped perspiring. Sometimes he would fall asleep. Then he would get up, burn perfumes to fumigate his body and have dinner brought in.

Despite this lifestyle of leisure and luxury, he found time to produce a staggering amount of work. He also turned translation into something of a family business, and both his nephew Hubaysh and his son Ishaq became important translators.

Qusta and Thabit

Besides Hunayn and his family, other famous Abbasid translators included Qusta ibn-Luqa, whose name means Constantine, son of Luke. Another was Thabit ibn-Qurra the Sabian. The Sabian community were polytheists who lived on the borders of what is now Turkey, and most of them were thoroughly at home in Greek, Syriac and Arabic, which is why many became translators. Thabit, though, was just a young money-changer in the small town of Harran until one of the Banu Musa brothers spotted his talent on the way back from a book-hunting trip to Byzantium.

Thabit's background in money-changing clearly stood him in good stead and he became famous for his translations of mathematical and astronomical works from Greek. As with so many of the Arabic translators, though, he did not simply translate. The translations were a starting point for his own ideas. One of the famous mathematical problems that Thabit is linked to is the chessboard problem, an example of an exponential series.

The problem goes like this. The man who invented chess so pleased his king that the king asked him what his reward should be. The man replied that he wanted nothing more than to receive one grain of wheat for the first square of the chessboard, two on the second, four on the third, eight on the fourth and so on, doubling until all 64 squares were filled up. The king seemed happy at such an apparently modest request. But of course the doubling actually means that the total number of grains is gigantic. Mathematicians loved this problem, and al-Biruni later calculated the answer as 18,446,744,073,709,551,615 grains (or 18.5 million trillion grains).

Al-Kindi

The doyen of the early Abbasid translation world was Ya'qub ibn-Ishaq al-Kindi. Al-Kindi was not a translator himself, but he was the head of a major translation circle working for the caliph. The 10th-century historian ibn al-Nadim recounted how al-Kindi was called the

Philosopher of the Arabs, 'unique in his knowledge of all the ancient sciences'.

Al-Kindi was a Muslim, but he also spent much time confronting the problems of reconciling faith and reason and providing a philosophical basis for Islamic intellectual life. Today we would call him a rationalist, and he wrote treatises exposing what he considered to be the charlatan nature of both astrology and its predictive powers and alchemy with its promises of turning base metal to gold. His belief in the power of logic and his willingness to search for answers everywhere, including Greek texts, later got him into trouble with Baghdad's rulers, but he was always held in high esteem in later centuries. He wrote:

> We ought not to be embarrassed about appreciating the truth and obtaining it wherever it comes from, even if it comes from races distant and nations different from us. Nothing should be dearer to the seeker of truth than the truth itself, and there is no deterioration of the truth nor belittling of one who speaks it or conveys it.

Yet accounts from the time tell us that he was not an easy man to get on with. He had a notoriously short temper, and the author al-Jahiz lampoons him in his *Book of Misers*. One story al-Jahiz tells is about one of al-Kindi's tenants who was foolish enough to ask him if he could have a guest to stay. Immediately, al-Kindi raised the rent by a third. It could have been such lofty disdain that prompted the Banu Musa brothers to arrange for the confiscation of his personal library, though he did get it back later. Al-Kindi is said to have been physically assaulted by people he had offended.

The original polymath

Al-Kindi was Arabic-speaking, from a noble family who settled in Kufa after the conquest. He is believed to have written many books, though it's not just the quantity that's impressive but the range too. He was the original polymath and seemed to write on just about everything, from astronomy to zoology. Most of the scholars of the time had wide-ranging interests, switching easily from science to philosophy to poetry, but al-Kindi's range was seemingly boundless.

Some of his interests were clearly driven by the needs of the caliphate. He wrote a famous treatise on metallurgy and sword-making. He also wrote on cryptography and described frequency analysis as a way to crack ciphers, which must have proved invaluable to the caliph's spies. Here is his beautifully simple explanation:

One way to solve an encrypted message, if we know its language, is to find a different plaintext of the same language long enough to fill one sheet or so, and then we count the occurrences of each letter. We call the most frequently occurring letter the 'first', the next most frequently occurring letter the 'second', the following most occurring the 'third', and so on, until we count for all the different letters in the plaintext sample.

Then we look at the cipher text we want to solve, and also classify its symbols. We find the most frequently occurring symbol and change it to the form of the 'first' letter of the plaintext sample, the next most common symbol is changed to the form of the 'second' letter, and so on, until we account for all the symbols of the cryptogram we want to solve.

Some of his interests, though, were more everyday. He was one of the first great perfumiers, coming up with some basic recipes and production techniques that are still sometimes used today.

One fascinating piece of text suggests that he was even thinking about time, space and relative movement – the very issues that modern physicists are still grappling with well over a thousand years later. 'Time exists only with motion,' al-Kindi says, 'body with motion, motion with body ... if there is motion there is necessarily body; if there is a body, there is necessarily motion.' He also used an Arabic word for 'relativity'.

The role of paper

One thing that arrived in Baghdad just in time to really help the translation movement, and the whole of Arabic scholarship, was paper. There is an apocryphal story that the Muslims learned the art of papermaking from Chinese prisoners they caught at the Battle of Tallas in 751. It's probably just as likely that paper arrived from China with the many traders who were at that time journeying far across Asia, and that they brought back Chinese calligraphy as well as paper. Either way, it arrived in Islam just about the same time as the founding of Baghdad by the Abbasids. Its impact was enormous. Parchment was very expensive, hard to come by, thick and awkward to use. Paper, on the other hand, was cheap, available in bulk, light and thin, and was perfect for a new calligraphic style of Arabic writing. If in China, papermaking might have been an art, in Baghdad it became an industry.

With paper, books could be made and copied comparatively cheaply in large numbers, and the boost this gave to learning in Islam is immeasurable. Previously, parchment codexes and scrolls of books had been so rare and so bulky and precious that they were held only in a very few private or royal libraries. With the coming of paper, books and bookshops appeared not just in Baghdad but in many other Islamic cities too. Even those who were moderately wealthy could build up their own private library, and public libraries appeared for the first time. In Bukhara, for instance, there was a public library where scholars could simply drop in, ask the librarian to get them a particular book from the library stacks off to the sides of the main hall, and then sit down to make notes. The library even provided free paper for the scholars. By the 13th century, Baghdad had many public libraries and bookshops, with numerous publishers employing scores of copyists to make the books.

It's hard to know whether the coming of paper stimulated the demand for books, or whether paper arrived because of the demand for books. Either way, it meant that with so many books and so many translations, the Abbasid scholars were very widely-read.

Beyond translation

The translation movement went on for more than two centuries under the Abbasid caliphs, and then seemed to peter out. It was partly because there were fewer interesting texts left to translate, but more because they no longer had anything to teach the scholars of the Islamic world. Almost as soon as they began the translation process, they had begun to think about what they were reading, and to make their own contribution. By the 10th century, they had much less to learn from the ancients. As the following chapters will show, there were, in particular, scientific developments, not just in Baghdad but across the empire. Achievements from Jabir ibn-Hayyan (Geber) in chemistry, Musa al-Khwarizmi in mathematics, and Abu Bakr al-Razi (Rhazes) in medicine stand out, but there were many others across a wide range.

5

The Caliph of Science

Knowledge has no borders, wisdom has no race or nationality.
To block out ideas is to block out the kingdom of God.

Aristotle speaks to al-Mamun in the caliph's legendary dream

In the traditional view of Islamic science, the start, and maybe the highest moment, of the Golden Age is the twenty-year rule of the Caliph al-Mamun, who ruled from 813 to 833, dying at the age of 47 while on campaign against Byzantium.

Al-Mamun was one of Harun al-Rashid's two sons, and he became caliph only after a violent civil war against his brother al-Amin. Al-Amin was the designated heir to the caliphate, but in a repeat of the Abbasids' coming to power, al-Mamun regarded himself as more deserving of the highest office and fought his brother all the way to Baghdad. The city was then subjected to a year-long siege that has been described as a medieval Stalingrad, drawing in not just armies but the city's inhabitants in vicious streetfighting. Finally, al-Mamun triumphed and al-Amin was killed. But the death of Amin was by no means the end, as other rivals appeared, and for six years al-Mamun attempted to rule from Merv, moving to Baghdad only in 819. Even then he faced a great deal of opposition in the west, and for the remaining fourteen years of his life he spent much of his time engaged in battles with opposition inside the Islamic empire, and against the Byzantines.

Dreaming of Aristotle

Alongside a lust for power, al-Mamun's rule was also characterised as a time when science and scholarship were at their peak. Al-Mamun is regarded by historians as the great champion of rationalism, and as the caliph who promoted science more than any other. It is said that once, when al-Mamun achieved a victory over the Byzantines, he asked from them as reparation not gold nor any other such mundane treasures, but a copy of Ptolemy's great book on astronomy, the *Almagest*.

There is a famous story telling how al-Mamun once saw Aristotle in a dream. Several versions of the story exist. Here is one transcript of the exchange:

Al-Mamun to Aristotle: What is good?
Aristotle: That which is in the mind.
Al-Mamun: What more is good?
Aristotle: That which is in the law.
Al-Mamun: What more?
Aristotle: The will of the people.

Al-Mamun: And what more?

Aristotle: There is no more.

In another more elaborate version, Aristotle explains that reason and revelation are not in opposition – that Man should seek God's truth by opening his mind to the power of reason rather than by waiting for divine revelation. He then goes on to instruct al-Mamun to turn all resources to translating the great works of thought and knowledge into Arabic, for 'Knowledge has no borders, wisdom has no race or nationality. To block out ideas is to block out the kingdom of God.'

The story then goes on to tell how, on waking, al-Mamun instructs men to go to Byzantium and bring back all the greatest books, to go to Gundeshapur in Persia and bring back the contents of its great library, to find all the best scholars and translators, and finally to build a centre at the court in Baghdad for learning and scholarship which he will call the House of Wisdom.

The House of Wisdom

Much attention has been paid to al-Mamun's House of Wisdom. Some researchers describe it as an institution for studying science and philosophy. It was here, according to such a view, that all the greatest scholars worked and debated, where there was the non-stop whisper of pens on paper as the great classical works were translated into Arabic. In this version, it was both a visionary scientific research institute and a proto-university. In fact, little is known about the House of Wisdom, and many historians now think its status as a university or research centre has been overplayed. It was almost certainly a library of books, and also a place for some translation as well as some astronomy, especially in al-Mamun's last years. But beyond that, the evidence from manuscripts is not solid enough to know more.

Still, it is clear that al-Mamun's interest in science was deep and genuine. Besides the House of Wisdom, he set up one of Islam's first observatories at Shamsiya in 829, and from the start it was making key updates to ancient astronomy such as the measurement of the solar apogee and the motions of the planets. He also had a map of the world drawn with as much accuracy as possible with the current state of knowledge. And in the 820s he instructed the Banu Musa brothers to check something that he had read in one of the ancient science books recently translated. This was a measurement of the circumference of the earth, given as 24,000 miles. With great ingenuity, the Banu Musa made their calculations and confirmed the accuracy of the ancient measurement, as we shall see later. Yet al-Mamun was still not satisfied. He sent the Banu Musa off to repeat the exercise in another place, and only when this too proved the ancient measurement correct was he satisfied. With a caliph this interested in scientific precision, it's hardly surprising that advances in science took off in such strides.

Helping on al-Mamun's great earth measurement project was one of the greatest of all the Muslim scientists, the brilliant al-Khwarizmi, and it was under the patronage of al-Mamun that al-Khwarizmi did most of his best work. Some sources suggest that al-Khwarizmi was attached to the House of Wisdom; others say that he worked independently. Either way, al-Mamun's Baghdad was the perfect setting for his talents to flourish. When he arrived there, he would have found Hunayn ibn-Ishaq translating Euclid's *Elements*, others translating Pythagoras, still more translating Archimedes' work on spheres and circles, and much, much more. Moreover, he was given the resources to trace key manuscripts all the way to India. Perhaps nowhere in the ancient world but Baghdad and at no other time but al-Mamun's would al-Khwarizmi have been able to fulfil his potential so spectacularly, as we'll see in later chapters.

A reasonable ruler?

There seems little doubt that al-Mamun played a major role in creating an encouraging setting for science in Baghdad and driving the translation movement. He championed scholarship for the sake of knowledge and for pragmatic political reasons. After all, the empire was still not entirely finished with wars, both at home and against Byzantium. The Abbasids were connected to the Prophet's family, but al-Mamum had just killed his brother and was keen to emphasise that he was the rational choice as caliph and linked to the divinely chosen family. Interestingly, he made a point of saying that Ali, Muhammad's cousin, was the best human being after the Prophet. So he was treading a careful line between opposing factions.

Launching a war on Byzantium in 830 may also have been a useful way of proving that he was a committed Muslim. On the way back from one campaign, he had the inscription changed on the Dome of the Rock in Jerusalem to suggest that he, and not the Umayyad Caliph Abd al-Malik, was the builder. Probably few were fooled, but it shows that public perceptions were high in al-Mamun's mind.

Al-Mamun's PR, if it was his, worked with many who told stories about him. He was portrayed as the arch-rationalist, the driving force behind the modernisation of Islam and the great champion of science and scholarship. He listened to a group of scholars called the Mutazilites. Later, the Mutazilites came to be seen as heretics by many Muslims, but their views may have made sense to al-Mamun because he wanted to build a powerful state based on reason.

The Mutazilites believed, like all Muslims, that the Qur'an is God's eternal word. However, in their view, it was created under God's guidance, and had not existed forever. They believed, too, that human reason is the key to wisdom and understanding God. This idea came partly from the input of Greek philosophy, but it was also a strand that had long been present in Islamic thinking, with its concept of *ilm* (knowledge) and *aql* (human intelligence and reason).

The Rationalist Inquisition

Yet, ironically, for all their defence of reason, the Mutazilites and al-Mamun forced their beliefs on others in an unreasonable way. The theological debates and battles that went on at this time have been far too simply characterised as a battle between the rationalists led by al-Mamun and traditionalists who believed the Qur'an was not created by man but spoken by God. Yet the so-called traditionalists, in the long run, turned out to be in some ways the more radical, or at least seemed to be fighting against an authoritarian status quo. Al-Mamun resorted to defending his position, and suppressing the growing opposition to it, with his own thought police or Inquisition called the *mihna*.

In the last year of his life, al-Mamun ordered that the governors of each of his provinces round up scholars to confess that the Qur'an had been created, not spoken by God. Those who refused were dismissed from public office, put in jail, and even flogged. Many who would otherwise have agreed with al-Mamun refused to confess, as they regarded the whole affair as the unwarranted interference of the state in personal affairs. In protest, some scholars would play mind games highly reminiscent of the interrogations of the Stalinist years, as the following exchange between the governor of Baghdad and a scholar-jurist called Bishr shows.

Governor to Bishr: What do you say about the Qur'an?

Bishr: It's the speech of God.

Governor: That was not my question. Is it created?

Bishr: God is the creator of everything.

Governor: Isn't the Qur'an such a thing?

Bishr: Yes.

Governor: So it's created?

Bishr: It isn't the same as a creator.

Governor: That's not what I am asking. Is it created?

Bishr: I have nothing more to say.

(From *Al Ma'mun* by Michael Cooperson,
Oneworld, 2005)

When word of such exchanges reached al-Mamun, he gave orders to behead dissenters. Under this threat, most climbed down. But it's hardly surprising that many Muslims now see al-Mamun not as the champion of reason and the initiator of Islam's Golden Age of Science but as an irreligious dictator who curtailed free speech.

The resistance

Perhaps al-Mamun's biggest mistake, in retrospect, was his persecution of Ahmad ibn-Hanbal, a major figure in Islamic theology and the founder of Islam's fourth school of law. Ibn-Hanbal believed that the caliph should have authority in political matters but not spiritual matters. This was a direct challenge to al-Mamun, who also saw ibn-Hanbal as a threat to science and rationalism and believed that nothing less than force would be needed to sideline ibn-Hanbal's views.

In 833, ibn-Hanbal was summoned to appear before al-Mamun, but the caliph died before ibn-Hanbal could reach him and he was brought instead before al-Mamun's successor, his brother al-Mutasim. The new caliph asked ibn-Hanbal to repeat what had been asked of him: that the Qur'an had been created by God. Ibn-Hanbal replied that such theological arguments were divisive, and that it was better for all if everyone agreed that the Qur'an was God's word and left it at that. The caliph would have none of this, and ibn-Hanbal was flogged. He wouldn't budge, and was taken to prison, where he spent the next 28 months. On his release, he was placed under house arrest.

The opposition to the rationalists now had its hero and martyr. Many years later, in his book *Heirs of the Prophets*, Rajab al-Hanbali, a follower of ibn-Hanbal's school of thinking, described those scholars – including the great scientists – who went with al-Mamun as being corrupt and beholden to the state. In contrast, those like Ahmad ibn-Hanbal were described as honest and righteous. Science and learning were therefore seen as synonymous with cruelty and dictatorship. Such a perception, however, would not be confined to the citizens of the Abbasid caliphs.

The pyramid quest

It's hard at this distance in time to see which picture of al-Mamun is more accurate – the enlightened champion of reason and hero of Islamic science, or the power-obsessed, irreligious dictator. It's likely that truths can be found in both. There is one story about him which seems to typify either his hunger for knowledge or his hunger for the power that goes with knowledge.

Apparently, word reached al-Mamun that the Great Pyramid of Giza contained accurate maps and charts of the earth and the stars. So in 820, he embarked on an expedition to Egypt with a team of engineers and scientists. For days, they scoured the smooth northern slope of the pyramid for a way in. Unable to find one, al-Mamun had his team hammer away at likely places. Making little or no progress, they tried heating the rocks with large fires and pouring cold vinegar on the hot limestones. Eventually, some of the masonry cracked, and al-Mamun's team found a way in. Still, plug after plug barred their way, and it must have required enormous determination to keep on going. At last, though, they found a chamber, now thought to be the Queen's chamber, but the chamber was empty … Or was it?

6

The Flowering of Andalusia

A palm tree stands in the middle of Rusafah
Born in the west, far from the land of palms
I said to it: How like me you are, far away and in exile
In long separation from family and friends,
You have sprung from soil in which you are a stranger
And I like you, am far from home

Poem written in Cordoba by the exiled Umayyad prince Abd
al-Rahman, founder of the Umayyad dynasty in Andalusia

Although it may not have seemed that way at the time,
the Abbasid revolution in 750 was to contribute over the
centuries to a deepening chasm in the Islamic world,
between east and west, and between Shia and Sunni
Muslims. It was no doubt to avoid any future challenge
to their rule that the victorious Abbasids invited the
Umayyad family to a reconciliation dinner in Damascus,
then slaughtered every single one of them. Or so they
thought.

Remarkably, out of that night of unimaginable
bloodshed was to emerge one of the greatest and most

unexpected triumphs of the Islamic revolution. Hidden away in the palace as the slaughter went on were the two teenage Umayyad princes Abd al-Rahman and Yahya. With the aid of their faithful Greek servant Bedr, the pair escaped before the alarm was raised. The Abbasids caught up with them as they swam for their lives across the Euphrates river. Yahya was beaten by the current and driven back into the hands of the Abbasid soldiers, who directly beheaded him. Alone, sixteen-year-old Abd al-Rahman and Bedr made it across, and began an adventure that took them through Egypt and across North Africa and into Spain, dodging Abbasid soldiers all the way. But it was not Abd al-Rahman's escape that was to prove the most extraordinary part of his story.

Islam comes to Iberia

Muslims had already set foot in Spain some fifteen years before young Abd al-Rahman was born, when a freed slave turned general called Tariq went to southern Spain to fight the Christian Visigothic King Roderick. Tariq's men won battle after battle against the much larger Visigothic armies – partly since many of the Visigoths switched over to Tariq. A great rock was later named after him, Jebel al-Tariq (Gibraltar, or Tariq's mountain).

By the time Abd al-Rahman reached North Africa, many parts of southern Spain, or al-Andalus as they called it, were being ruled by Muslims. But there was considerable discontent, not so much, ironically, among the Christians as among some of the Muslims who had helped achieve victory but were now being sidelined by rulers as they carved out their own domains and city states. This was the situation that the young Umayyad prince would encounter and take advantage of as he sailed across the Straits of Gibraltar to Andalusia.

The ornament of the world

Small numbers of Muslims joined Abd al-Rahman as he progressed through Spain. He was able to defeat the leader of Cordoba, then one of Spain's principal cities, allowing him to declare himself ruler in his place. Over the next decade, with remarkably adept leadership, Abd al-Rahman, who became known as the Falcon of Andalus, extended his influence.

Yet he was not content to simply gain control. He wanted to restore the glory of the Umayyads, and his way of doing that was to recreate Damascus in faraway Spain. When he arrived, Cordoba was important but possibly run-down. Abd al-Rahman created a city that turned out to be a focus of culture, learning and science to rival Abbasid Baghdad. Umayyad Cordoba was to have its own Golden Age of learning.

At the court of the Holy Roman Emperor Otto I, the nun Hroswitha of Gandersheim wrote in 955 that

The brilliant ornament of the world shone in the west, a noble city newly known for its military prowess that its Hispanic colonisers had brought, Cordoba was its name and it was wealthy and famous and known for its pleasure and resplendent in all things, and especially for its seven streams of wisdom ...

(Quoted in *Ornament of the World* by Maria Rosa Menocal, Little, Brown, 2002)

At Cordoba's heart was a new mosque that was ultimately to become one of Islam's most magnificent, but the city was said to have hundreds. It is also said to have had many bath-houses and hospitals. For himself, Abd al-Rahman built a palace, nostalgically named the Damascus Palace, in the midst of beautiful gardens, created to recall his grandfather's country retreat in Syria, complete with palms and many other plants seen in Spain for the first time – hence the poem at the head of this chapter.

Beyond the gardens, elegant villas, courtyards and fountains, green spaces and broad paved streets made Cordoba one of the most desirable places to live in all of Islam, and it soon attracted a wealth of talent. Inventive minds and ample cash ensured that the city had access to modern amenities, from running water in every substantial house to street-lights. Meanwhile, the introduction of the latest farming techniques and irrigation methods turned the countryside around into productive farmland, with vast new orange and olive groves as well as cereal fields, to ensure that the city stayed well fed.

Cordoba's jewel, though, was its library. The Umayyad library in Cordoba was just one of more than 75 in the city, yet it is estimated to have contained 400,000 books – at a time when the largest library in Europe held much less. Edward Gibbon underlined its huge scale by pointing out that the catalogue for the library alone ran to 44 volumes. A lot of books have survived from this period of Islamic rule, but this vast number suggests that it is only a small fraction of the books in circulation at the time, the rest burned or lost as the Umayyads' power in Spain waned. It is tantalising to think what might have been in these books, and what contributions to science they might have made. Perhaps the most important books were physically saved, or carried away to safety in the minds of leading thinkers like Musa bin Maymun (Maimonides) and Walid ibn-Rushd (Averroes). But there is no way of knowing.

Attracting talent

For centuries under the Umayyads (who finally declared themselves caliphs in opposition to the Abbasids in Baghdad in 929) Cordoba shone, and many of Islam's brightest and best began to think of Cordoba and Andalusia as the place to be rather than Baghdad. There was both money and a very attractive lifestyle.

One of the talents drawn here was the extraordinary Abu al-Hasan Ali ibn-Nafi, usually just known as Ziryab, which means 'blackbird'. Originally a slave from Iraq, Ziryab was brought to Cordoba in 852 by the ruler Abd al-Rahman III with a handsome salary. Very soon, the blackbird excited people with his new five-string version of the *oud* (an early version of the Spanish guitar) and with his love songs. People cut their hair short like him.

They followed his fashion for vividly coloured cotton and silks in summer and richly hued woollens in winter. He introduced them to fine porcelain tableware, simple three-course meals, playing chess and polo, and even toothpaste. Indeed, it seemed there was barely an aspect of leisured life for which Ziryab did not set a striking new fashion.

Yet it was not just for fine living that many talented people came to Andalusia; it was a magnet for scholars, too – and particularly thinkers and philosophers. The Umayyad leaders and their wealthy entourage ensured that Cordoba and other Andalusian cities such as Toledo and Seville all had well-stocked libraries and well-paid jobs for intellectuals. Like Baghdad, Cordoba was a cosmopolitan city, and Christians and especially Jews found that they were welcome here. For many centuries, Jewish intellectual life flourished in what has been called the Jewish Golden Age, and many Jewish thinkers were involved in what is called the Toledo School, translating works of Arabic into Latin.

In addition to Muslims, other scholars were Christians, or Mozarabs – Christians who had learned to speak Arabic and who had adopted Arab lifestyles. Mozarab scholars were later to play a key role in the transmission of the new knowledge from Arabic into other parts of Europe. They were invaluable conduits, just as the Nestorian Christians had been, in the translation of classical knowledge to Arabic in Baghdad.

The presence of different religious communities in Spain, later called *convivencia*, is sometimes compared with today's multi-cultural societies in Western countries. There were undoubtedly some similarities with the present. For example, just as today, science, learning and innovation in those earlier centuries was the product of researchers and scholars from different nations and different cultural backgrounds working together, or sharing the same space. But there were also differences. Christians and Jews in al-Andalus had a different status in law compared with Muslims. As was the case elsewhere in the Islamic world, non-Muslims paid a different set of taxes, they were exempt from military service, and they were not entitled to be considered for the post of ruler or caliph. Moreover, there were restrictions on preaching religions other than Islam, or speaking ill of notable figures from Islamic history. As a result, in the 9th century, a large group of Christians killed themselves in protest at not being allowed to criticise the Prophet, which Muslims regard as blasphemy.

Wings of fire

One of the first of the Andalusian scientists of this period was Abbas ibn-Firnas. Born in Izn-Rand Onda (modern Ronda) in 810, he was originally brought to Cordoba to teach music. But once there he began to show his extraordinary range of talents as an inventor. He had a special interest in glass, and is claimed to have produced clear drinking glasses. Colourless, transparent glass had been around since Roman times, but by manipulating the mix, he created glass so clear that the contemporary poet al-Buhturi said it was as if the contents were standing there without the container. Presumably such clear glass would also have been useful for the lenses ibn-Firnas made to correct sight and to magnify things. One of his most talked-about inventions was a sky simulation room. Inside the room, there was not only a giant machine that showed how the planets moved, but people were also astonished to see stars, clouds, thunder and even lightning produced by hidden mechanisms in the inventor's basement.

Most famously, though, ibn-Firnas is said to have been among the early pioneers of flight. Some sources say he was inspired by seeing in 852 a stuntman called Armen Firman surviving a leap from the top of one of the minarets near the Grand Mosque wearing a very loose silk cloak stiffened with wooden struts. Others say that the stuntman was actually ibn-Firnas himself, and Armen Firman is simply a latinised form of his name, which seems likely. Survivor or onlooker, ibn-Firnas decided he could do better. For the next attempt, he constructed a large wing like a modern hang-glider from silk and eagle feathers on a light wooden frame. After a few successful trial runs out in the desert, he decided it was time to make a public attempt.

A large crowd gathered as ibn-Firnas, now nearly 70, strapped on his wings at the top of a high cliff in Cordoba's Rustafa garden. Then, with amazing courage, he leaped into the void. To everyone's astonishment, he did not plummet straight down. Instead, he floated out in the air, circled a few times for up to ten minutes, then gradually came in to land. Unfortunately, he hadn't realised how much he would need to slow down, and he crashed into the ground, breaking the wing and badly injuring his back. Later, he realised that birds use their tails to slow down when landing – and he had crashed because he didn't have one. But he was too old now to try again. Or so the story goes. As with so much of science from the Islamic era, our knowledge of ibn-Firnas comes from third-party reports, rather than the man himself.

Surveying time

Nothing quite matched ibn-Firnas's flight for sheer spectacle, but some of the devices made by al-Zarqali (Arzachel) in Toledo in the 12th century were even more ingenious. Al-Zarqali started off as a simple metalworker, but he was apparently so skilful in making astronomical devices that Toledo's astronomers urged him to learn more about the theory of astronomy. After studying for several years, al-Zarqali began to design his own astronomical devices.

In particular, he came up with a sophisticated new astrolabe design which set the standard for centuries to come. There were very down-to-earth practical uses for this elaborate astronomical device. A later Andalusian scientist, Maslama al-Majriti (of Madrid), found that he could use the astrolabe for surveying fields accurately, so sorting out some of the complicated inheritance problems that often beset landowners. Al-Zarqali also came up with an easy-to-use astrolabe designed for someone with little astronomical knowledge. His model dispensed with the complex 'spider' which showed the movement of the planets, and so was easy for a beginner to master. Such junior astrolabes were used in astronomy, surveying, and also in astrology.

Al-Zarqali's pièce de résistance was an ingenious water clock that told not only the hours of the day but the phases of the moon – important in Islam for working out the beginning of the new lunar month. This device was a tourist attraction in Toledo for many centuries to come, until a curious inventor was given permission to take it apart to see how it worked – and couldn't put it back together.

Al-Zarqali's work wasn't simply practical, though. He made a number of key contributions to astronomy. He proved that the aphelion – the point at which the distance between the sun and earth is greatest – shifts very slightly each year against the fixed background of the stars. He measured this minuscule movement, and worked out that it was shifting at 12.04 seconds each year. Modern measurements put it at 11.8 seconds. His accurate measurements also helped him to contribute to the astronomical *Tables of Toledo*,[1] renowned for their accuracy

---

[1] The *Tables of Toledo* were the work of a group of Toledan astronomers in the 11th and 12th centuries, led by Sa'id al-Andalusi and including al-Zarqali. They were mostly based on existing tables by al-Khwarizmi and al-Battani, but included some new observations by al-Zarqali. It is thought that they were essentially an attempt to adjust data to the latitude of Toledo, which became the new astronomical meridian. But for all their care, they apparently contained serious errors in relation to the movements of Mercury and Mars.

and later quoted by Copernicus. And he created the first *al-manakh* (almanac), which contained tables that allowed people to compare the Islamic calendars with others very simply and easily for the first time.

Doctors of Spain

The Prophet's injunction to care for the sick meant that medical science, too, flourished in the Islamic era, and Cordoba, like other cities, became renowned for its healthcare. Cordoba had hospitals with running water, baths, separate wards for different ailments each headed by a specialist, and a requirement to be open 24 hours a day to anyone who was sick. There were also numerous physicians, many of whom trained at the hospitals of Baghdad. When they came back to Spain, some were lucky enough to be housed in the palace complex outside Cordoba at Medinat al-Zahra (the city of Zahra, named after Muhammad's daughter).

One of the most popular Andalusian doctors was ibn-Shuhaid. He recommended that his patients improve their diet, and treated them with drugs only if this proved ineffective. The majority of the Andalusian doctors were very likely to be good practical physicians rather than groundbreaking medical theorists. Al-Zahrawi, who died in 1013, became one of the most famous surgeons in

---

Nevertheless, they became famous throughout Europe over the next few centuries, and in Chaucer's *The Franklin's Tale*, the clerk proudly displays his 'Tables Tolletanes'.

Europe, and his manual of surgery became a standard textbook for centuries. Meanwhile, ibn-Zuhr (Avenzoar) wrote *al-Taysir* around 1150 for Almohad Caliph Abd al-Mu'min. It became a guide to therapeutic treatments in universities.

New thinking

Undoubtedly, philosophy is among the strongest legacies from Islamic Spain. Three names in particular stand out: these are ibn-Rushd (Averroes), ibn-Arabi and Musa bin Maymun (Maimonides, a leading thinker in Judaism). Maimonides and ibn-Rushd worked in many fields, but all three asked fundamental questions about the nature of religion and science.

As we'll see later, ibn-Rushd wrote a famous treatise, the *Incoherence of the Incoherence*, to rebut al-Ghazali's polemic against science in his *Incoherence of the Philosophers*. Al-Ghazali had asserted that conclusions reached by reason and human intellect alone were not enough to understand the complexities of the natural world and, moreover, were incompatible with the teachings of the Prophet.

Ibn-Rushd countered that the Qur'an instructed humans to look for knowledge, so the search must at least be right. He believed revelation to be the highest form of knowledge, but felt that the majority of people were ill-equipped to grasp the complexities of the religious experience and therefore needed something simpler, which was a theology based on human reason.

Ibn-Rushd had a significant influence on thinkers in the Latin world, such as Thomas Aquinas.

The year 1165 saw the birth of Muhammad ibn-Arabi in Murcia in the south of Spain (he died in 1240 in Damascus). Ibn-Arabi positively revelled in letting his imagination roam far and wide and scale new heights. He was confident in his abilities, and set about thinking how to create a theory of everything. At the same time, he also attracted charges of heresy from some in the Muslim mainstream. This may have been because ibn-Arabi believed that (apart from the Prophet Muhammad), God had endowed special powers on a select number of individuals and that he was one of them.

Ibn-Arabi believed that human reason is both useful and powerful, in explaining the world and in creating great works of learning. But for him, human reason is but one of many components in what constitutes 'knowledge', and he would point out something that even the best scientists are well aware of: that sometimes in science, the answers to difficult and seemingly intractable problems, or even 'breakthroughs', happen for reasons that cannot be explained by 'reason' alone. Scientists often talk about being inspired, or how sudden moments of inspiration helped them to make progress. Is a flash of inspiration an example of human reason, ibn-Arabi asked? For him, there was clearly some other unexplained force that worked alongside reason. Harnessing that power would be the key to 'knowledge'.

Ibn-Arabi focused his energies into an idea which later historians have called the 'unity of existence' – and

which stands as his greatest legacy. The unity of existence says that everything that exists in the universe is
connected together and also to God. Ibn-Arabi was not
the clearest of writers, which has meant that ideas such
as this have been interpreted in different ways, or completely misinterpreted. Did ibn-Arabi mean that God,
the planets, humans, plant and animal species all come
from the same material? Or is he saying that they have
characteristics in common? Is all of creation a kind of
giant and complex super-organism? Or taken another
way, does he mean that humanity and the wider environment are like a large family and that the actions of one
individual will have consequences for others? Yet another
interpretation of the unity of existence is that there can
be no outsiders or 'others', which could make ibn-Arabi
an early anti-racist.

Final flowering

By the 11th century Umayyad control was waning, and
in 1090, eight years before ibn-Rushd's death, Cordoba
was captured by the Almoravids. This is a Spanish name
for a group of Muslims from North Africa known in
Arabic as *al-Murabitun*, or those who stand together for
the defence of the faith. They were distinguished by a
face muffler, which they wore rather like the Tuareg, to
whom they are distantly related. The Almoravids were
against what they saw as the luxury and decadence of
the Umayyad courts, and they conquered in the name of

returning Andalusia to Islam's fundamentals – a development that has happened so often in Islamic history.

With the fall of the Umayyad caliphs in Cordoba, al-Andalus split into a variety of city states. The Almoravids lasted a century and were followed by the Almohads, who ruled from 1130 to 1269. The Almohads were known in Arabic as *al-Muwa-hidun,* or 'the unitarians', and like ibn-Arabi they took the view that the universe and all of life are a part of God. Not unlike the Almoravids, they were also zealous about reforming what they saw as Spanish society's lax morals. They persecuted, equally, Muslims who disagreed with them and non-Muslims, which meant that people such as Maimonides were forced to leave home and settle in other parts of the Islamic empire. It is interesting that more than two centuries of an often violent and anti-intellectual atmosphere witnessed both the flowering and the decline of learning, and especially philosophy, in Islamic Spain.

Many other Andalusian scientists introduced earlier, such as ibn-Zuhr and al-Zarqali, were working around the same time. And this period further produced two of Islam's best-known geographers and travellers: al-Idrisi of Cordoba and ibn-Batuta. Al-Idrisi produced a famous map of the world for the Norman King Roger of Sicily in 1139, and also one of the first great geographies of the world, known as the *Book of Roger.* In this, he describes the climates, people and products of the entire known world. He also tantalisingly tells the tale of a Moroccan navigator who was blown off course in the Atlantic and sailed west for 30 days, before returning to tell of a fertile land far across the sea.

Ibn-Batuta of Granada was one of the great travellers of the age. In 1325, at the age of 21, he set off on the Hajj to Mecca. He finally came home after 24 years, having visited not just Arabia, but Egypt, Syria, Iraq, East Africa, India, Russia, and even China and Sumatra. The story of his journey is one of the best travel books of all time, and added considerably to people's knowledge of the world.

At the same time as the exodus of Muslims and Jews from al-Andalus had begun, Christian kings were gradually regaining control of Spain. Quite often, the new rulers were happy to respect Islamic ways as they moved south, and Arabic inscriptions were often incorporated in new churches and synagogues. But in 1492, the *reconquista* under Queen Isabella was finally complete, and Jews and Muslims had to leave al-Andalus permanently. Many settled across the waters in Morocco or in other Islamic lands – where many of their descendants still live. A little later the Spanish Inquisition began.

7

Beyond the Abbasids

To whom, indeed, can it be easy to write the announcement of the
death blow of Islam and the Muslims, or who is he on whom the
remembrance can weigh lightly? ... For even Antichrist will spare
such as follow him, though he destroy those who oppose him, but these
Mongols spared none, slaying women and men and children, ripping
open pregnant women and killing unborn babes.

Ibn al-Athir writing of the Mongol invasion of Persia in 1221

For 200 years or so under the Abbasid caliphs the Islamic
empire was vast and Baghdad was among the world's
wealthiest and most dynamic cities, a magnet for thinkers and scholars. Yet as with the Umayyads before them,
not all were happy with the Abbasids and there were
troubles on the fringes of the empire. Just as during the
Umayyads' twilight years in Damascus, discontent was
brewing.

Sensing weakness at the centre, governing families of
individual provinces began to establish their own territories independent of the caliphate. In Central Asia, the

Safavids and Samanids set themselves up as rulers, while out to the west, in North Africa, the Tulunids declared themselves in Egypt and the Aghlabids in Tunisia and then Sicily.

With less and less revenue flowing into Baghdad and the caliph's power dwindling, the irrigation systems in Iraq began to break down for lack of maintenance and agricultural production dropped. Baghdad was slipping into a spiral of decline. To maintain a semblance of power, the caliph was forced to rely upon his professional army – and so they gained more and more influence.

No longer feeling that the Abbasid caliph was in control, the Fatimids, a family that claimed descent from the Prophet Muhammad (through his daughter Fatima), declared their own caliphate in North Africa in 909. Twenty years later, the Umayyad leader in Cordoba, Abd al-Rahman III, declared himself caliph too. And finally, in 945, one family of generals, the Buyids, marched south to Baghdad from the Elborz mountains in the north of Iran and stormed the city. The Buyids left the caliph in power nominally, but adopted the title sultan and *shahanshah* (king of kings) for themselves, reminiscent of the old Sassanian kings.

The Fatimids

Among those who succeeded the Abbasids, it was the Fatimids who perhaps did most, as a caliphate, to continue advances in science and learning. The Fatimids had come to power on the basis that they had a relationship to the family of Muhammad and were therefore the legitimate heirs to the Prophet. They belonged to the Ismaili tradition, a branch of Shia Islam. The Abbasid administration was multi-cultural and multi-religious yet increasingly unpopular with Muslims, in part because of the interrogation of those who refused to agree with their rationalist take on religion. The ending of the Abbasid inquisition by the Caliph al-Mamun weakened them further and the close of the 9th century, therefore, saw pockets of dissent in many places, as well as, for the first time, the emergence of a power in the Fatimids who were strong enough to challenge the Abbasids.

At last, Muslims of the Shia tradition were in power. They had waited nearly three centuries for this moment, and when it came they ensured that it would last as long as possible. The Fatimids ruled from 909 to 1171, with Cairo as their capital. Like the Abbasids, the Fatimids were patrons of science, medicine, engineering and learning. They were rationalist to a degree, and could be as hard-core as, say, al-Mamun. The origins of the Fatimid caliphs lay in missionary work, and for them, learning was seen as an important vehicle for the spread of faith – not unlike the view taken by al-Mamun's opponents, the traditionalist scholars of Baghdad. In this and in other ways, they were closer to popular religion compared with the Abbasids, which meant, for example, that they had little time for astrology. They were keener on astronomy, however, and better builders of institutions. Their most famous legacy, Al-Azhar university, built in 988 initially to train Fatimid missionaries, is one of the world's oldest fully-functioning institutions for teaching and research.

If there is one Fatimid caliph who stands out above the others, a strong candidate would have to be al-Hakim, who ruled for 25 years from 996 to 1021. He was as much a patron of science as the Abbasid al-Mamun, but had less of the urge to conquer Christianity and quickly concluded an armistice with the emperor of Byzantium. Similarly, he seemed to have no desire to smother his people with his own personal take on religion, and was happier to see the Fatimid caliphate provide a neutral space for Islam's many (and rapidly dividing) traditions, from the extremely rationalist to the extremely orthodox. Famous scientists that were Ismaili (or brought up under the Ismaili influence of the Fatimids) included ibn-Sina, author of the *Canon of Medicine*, and the mathematician Hassan ibn al-Haitham, who made important discoveries in optics and in astronomy among other things.

An Islamic world

Within the space of less than half a century, as quickly as it had come into being, the vast Islamic empire that stretched 4,000 miles from Central Asia to the Atlantic had disintegrated. Or at least so it seemed. The remarkable thing is that that wasn't how it turned out at all – at least for a while.

In just three centuries, Islam had laid down such deep roots in nearly all the regions of the world it had touched that even with the power of the one caliph all but gone, and the empire entirely broken up, Islam was as strong as ever. There were now three separate caliphs – one in Baghdad, one in Egypt and one in Spain – and dozens of local rulers who paid them no heed. Yet the Islamic world was, if anything, even more unified in culture, language and religion than ever.

So as Baghdad's glory faded a little, other cities across Islam began to shine. Cordoba already rivalled Baghdad, and when the Umayyads lost their dominance in Spain, so the city states of Seville, Toledo and Granada burst into life. In Morocco, there was Fez. And out in the east, in Central Asia, the Safavids and Samanids established thriving cities – Tashkent, Samarkand and Bukhara – that emulated Baghdad in culture, learning and architecture. Another appeared in what is now Afghanstan at Kabul. The new city that made the most lasting impact, though, was al-Qahirah, or Cairo, established by the Fatimids near Fustat in Egypt in 969.

New centres

There was no shortage of opportunities for scholars and scientists in any one of these new alternatives. If anything there were more. But Islamic learning was no longer concentrated effectively in one place, and so it is much harder to see the thread of what was going on (which is why this continuation has only recently been appreciated by historians). Indeed, it must have been as hard for scholars at the time – hard to know where to go for the best opportunities, and hard to keep abreast of the latest developments in cities up to 4,000 miles away. Thus different centres began to go different ways.

Each of the centres acquired its own scientific stars in time. Seville in the 12th century, for instance, had the physician Abu Marwan ibn-Zuhr and the great astronomer Nur al-Din ibn-Ishaq al-Bitruji (Alpetragius). Al-Bitruji was not one of those who seriously questioned Ptolemy, but his mathematical rigour and calculations helped to provide some of the groundwork for Copernicus to create his heliocentric theory centuries later. In Madrid, there was the astronomer Maslama al-Majriti who perfected the astrolabe and gave Europe its first taste of Islamic astronomical tables. In Toledo, as we have already seen, there was al-Zarqali (Alzachel).

Also, as Christians were beginning to push south into Spain in the first stages of the reconquest, in the east came increasing incursions from the Turkic peoples. Most had converted to Islam but had a different culture to those who spoke Arabic. Over time, they edged further and further west into the heartland of the Abbasid Caliphate. In 1040, a group called the Seljuk Turks invaded Syria and Mesopotamia and, five years later, captured Baghdad.

As with the Buyids, the Seljuks called themselves sultans and let the Abbasid caliphs remain in Baghdad. There were still opportunities for scholars in the east, but they had to be more and more careful to attach themselves to the right patron and avoid being caught in the wrong place at the wrong time – like Omar Khayyam, who was caught in Isfahan after his patron, the Sultan

Malik Shah, was killed in battle and his other protector, Nizam al-Mulk,[1] was assassinated. Khayyam was lucky only to be forced to go on a pilgrimage to Mecca.

Heading for Cairo

For a while, Cairo must have seemed the safest bet – far from the troubled eastern and western fringes – and it attracted some of the greatest minds of the age, such as astronomer ibn-Yunus and physicist ibn al-Haitham. Cairo was one of the great medical centres, and eventually had three big hospitals: the ibn-Tulun, one of the first Islamic hospitals, dating to 872; the famous al-Mansuri where ibn al-Nafis was 'Chief of Physicians' in the 13th century; and the Qalawun hospital which was set up by Sultan Qalawun in 1284 and lasted some 650 years until it was demolished in the early 20th century. Part of the complex can still be seen today.

Hospitals existed before Islam, of course, but Islam's duty of care meant that early Islamic hospitals were often well funded by benefactors. The effect was to create

---

[1] Nizam al-Mulk (1018–92) is one of the most fascinating figures in medieval Islam. He was the advisor for the first two Seljuk sultans and acquired a reputation as an effective administrator of a huge empire to match the Barmakids – wise, prudent, resourceful and successful, and a devout Muslim. But he also created a legacy of efficient civil service which was to become a hallmark of later Turkish Islamic governments. He wrote down many of his ideas on how things should be run in a famous book called *Seyasat-nameh* (*The Book of Government*; or *Rules for Kings*). More importantly, he founded the *Nizamiyah* system of colleges.

hospitals that were models of healthcare for their age, and which seem surprisingly modern in their arrangements.

From the records, we know that the hospital was of a cruciform shape and divided into wards in which patients were separated depending on the type of illness they were suffering from. The mentally ill were kept apart from those with physical symptoms and men were housed separately from women. There were also separate units for patients with eye disorders, stomach complaints and those needing surgery. Hospital doctors by that stage had begun to specialise, and the Qalawun hospital's records tell us that it employed physicians, surgeons and opthalmologists, as well as administrators, nurses, accountants and orderlies. According to one report, many new patients were treated every day. If someone died during their stay in the hospital, the endowment would also pay for their burial expenses.

Cairo was also the location for Al-Azhar university, founded by the Fatimids. Historians debate the debt that European universities owe to these older Islamic universities, but they were definitely pioneers in the field of higher education. They were initially mosques designed for teaching, but the curriculum gradually extended to include a wide range of subjects. Students would travel long distances to study at Al-Azhar.

In 1005, the Fatimid Caliph al-Hakim gave Cairo its own *Dar al-Hikma* (House of Wisdom). It was essentially an academy for teaching astronomy, maths, medicine and astrology, but it also taught the Shia beliefs of its Fatimid patrons – which occasionally inspired the resentment of the predominantly Sunni population. Soon, Cairo acquired its own intellectual salons too, run by al-Hakim and his successors.

Ibn al-Haitham's madness

Al-Hakim is one of those figures over whom opinions are quite divided, and to this day writers from the Ismaili tradition say that he has been unfairly caricatured by his enemies. Traditionally, he has been described as an erratic and cruel ruler. Yet these descriptons come from sources who were opposed to the Ismailism that al-Hakim espoused.

One of the stories about al-Hakim is in relation to the physicist ibn al-Haitham. In Egypt, the annual flooding of the Nile was both a blessing and a curse. It brought the water for crops, and the rich silt in which they grew, but it also caused widespread devastation. The magnificent nilometers built by the Egyptians to check the levels of water in the river are a testament to how seriously and scientifically they took the problem. Still working in Basra, ibn al-Haitham came up with a plan to manage the Nile's floods by building a dam up-river. Al-Hakim heard of ibn al-Haitham's plan and brought him to Cairo to put it into action. Ibn al-Haitham arrived and set off up the river to get to work. By the time he had got as far as Aswan, he realised that the Nile was still far too wide to ever create a dam, and that his plan would never work.

Now, the story goes, he had to go back to the caliph and tell him that the plan failed, but fearing the wrath of the irascible al-Hakim, ibn al-Haitham feigned madness, acting even more strangely than the caliph himself. It was a risky idea, but ibn al-Haitham's mad ramblings and subsequent stupor apparently convinced the caliph and he was kept under effective house arrest in the Al-Azhar in Cairo. Here, the peace and quiet and the patronage of al-Hakim's sister Sitt al-Mulk gave him the scope to do his brilliant work on optics. There is no way of knowing how true the story is, but ibn al-Haitham's scientific achievements are indisputable.

Cairo was not immune to the divisions being experienced in Islam, but at the same time, different caliphates found temporary unity in the shape of opposition to the Crusades. In 1171, the famous Salah al-Din Ayyubi (Saladin) overthrew the Fatimids and reunited the forces of Islam sufficiently to drive the Crusaders out. Still, Cairo and Egypt remained relatively safe compared to what was happening in the east of the Islamic world.

The coming of the Mongols

In 1219, Genghis Khan and his Mongol army conquered China with swiftness and ease, and then turned westwards. With an army numbering up to 800,000, many with incomparable skill on horseback, the Mongols were unstoppable.

Bukhara was quickly overwhelmed, and when they reached the university town of Nishapur, the slaughter was dreadful. Men, women and children in the city were beheaded, then disembowelled. In other cities that fell in the Mongols' path, people were rounded up and slaughtered for sport like cattle. Those that survived the onslaught faced years of starvation, as the Mongols destroyed the *qanats*, the tunnels that supplied the fields with water.

When Genghis withdrew, the people of Islam might have sighed with relief. But the Mongols were not done. In 1256 and 1258, they were back, now led by Genghis's grandson Helagu. This time, even Baghdad was not safe. Ignoring the caliph's warning that his death would bring chaos to the world, the Mongols entered the city, killed the caliph and slaughtered hundreds of thousands of people.

No one knows how many citizens of the Islamic empire died at the hands of the Mongols, or how many died of starvation in their wake, but estimates suggest that it was many, many millions. One historian of the time suggested that the blow was so devastating that the population would take a thousand years to recover. He was not far wrong. The population of the region finally reached its pre-Mongol level again just a few decades ago.

The slaughter of those years is still etched deep on the memory of people in these parts. Yet there was still another dreadful invasion from the east to come in 1384. This was the Tartar army of Tamerlane the Great, which left a tower of heads piled in the square at Isfahan, and another in Baghdad, painfully rebuilt in the century after the Mongol destruction.

Survival

It would be easy to think that these unimaginable traumas would have spelled the death of Islamic science and culture. And yet, remarkably, they didn't. Helagu Khan, the destroyer of Baghdad, for instance, became a Muslim and a patron to one of the greatest Muslim astronomers, Nasir al-Din al-Tusi; while in Iran, when the slaughter stopped, Tamerlane and his Timurid successors presided over the continuing flowering of Iranian culture. Indeed, some of medieval Islam's greatest scientific achievements occurred after the Mongol and Tartar invasions.

Yet there is no doubt that these ghastly events had a deep and lasting effect on Muslim consciousness. The success of Islam and the towering intellectual achievements of the early years had seemed to confirm that God was on the side of Muslims. It was this self-belief – this conviction that the world was theirs to explore, to *know* – that enabled Islamic-era scientists to accomplish so much. The trauma of the destruction of Baghdad and so much else undermined that self-belief. And although there were still many individual scientific triumphs to come, many more important accomplishments in the field of knowledge, there was perhaps never the same drive and energy coming from every level of society, from the caliph down, that there had been in the first seven centuries of Islam.

Part II

Branches of Learning

8

The Best Gift From God

The best gift from God is good health. Everyone should reach that goal by preserving it for now and in the future.

Saying of the Prophet Muhammad on health

In few areas did science, community benefit and religious duty come so closely and productively together in early Islam as they did in medicine.

The Prophet Muhammad frequently emphasised the importance of good health and healthy eating. He also encouraged people to seek medical treatment, and is reported to have said: 'Make use of medical treatment, for God has not made a disease without appointing a remedy for it, with the exception of one disease – old age.' And by his emphasis on charitable works, one of Islam's foundational pillars, he encouraged doctors to provide care for the sick, and the wealthy to pay for it.

There were practical reasons for promoting medical science, too, of course, in the new empire. Battle wounds were all too common, for example, as were diseases of the digestive system and infections – transmitted, much as happens today, by the movement of people across borders, and into the new Islamic cities such as Baghdad. Yet empires had needed medical science before and had not provided it. What perhaps made Islam different at the time was the willingness of richer people to pay for healthcare, whether for religious, social or political reasons. Also important was the sense of religious duty that drew many people into the medical profession – as well as the excitement of the search for knowledge and the prospect of a lucrative career.

Of course, there were many charlatan doctors and quacks, but there were also several dedicated physicians. Some were at the very cutting edge of research and practice. Others just did the best they could with the tools at their disposal. The combined result is that the peoples of the Islamic empires perhaps had medical care which was as good as, if not better than, that of any empire that had gone before.

It would be inappropriate to compare the range and effectiveness of medical treatments in earlier times to what we have in the modern world. Nor did the Islamic world create the first hospitals. However, hospitals were advanced for their day, and doctors of the Islamic era often provided effective treatments. What is also clear is that the medicine that originated in Islamic times penetrated deep into Europe in the following centuries, perhaps more so than any other Islamic science. A whole range of books by physicians and surgeons such as Hunayn ibn-Ishaq, ibn-Sina and al-Zahrawi were in widespread use in the universities of Europe for many centuries. Their popularity only waned after the main basis of their theories, the idea of the four humours, was superseded by the germ theory of disease.

The Greek legacy

At the start of the Islamic era, various traditions of medicine were in use in addition to those of Arabia and, together, they served the growing cities of the empire. At Gundeshapur in Sassanian Persia, for instance, Persians and Nestorian Christian refugees from the Byzantine empire had founded a respected medical school. Some of these physicians moved to Damascus and Baghdad to establish elite medical dynasties there. But the biggest influence by far was Hellenistic – the medicine of the Greeks, which today is still practised in large parts of South Asia, and is known by the name *Unani*, which means 'Greek'.

Ancient Greek medical ideas became central to mainstream Islamic medicine, and when the translation movement got under way, there was a rush to translate as much of the work of the great Greek medical authorities as possible into Arabic.

One of the prize Greek medical sources was *De Materia Medica*, written in the 1st century by a Greek surgeon serving in the Roman army called Dioscorides. Dioscorides' book became the main source on drugs and on the herbs from which they could be extracted. But the most influential by far of the Greek physicians who worked for Rome was Galen, whose voluminous writings covered the entire field of medicine and provided a complete background in theory and practice for any physician.

Born in Pergamon in Turkey, Galen went to Rome as a young man, where his skills as a physician soon found him in the service of the emperor, and where he gained a reputation that made him the greatest authority in Western medicine for nearly 1,300 years.

As he was not allowed to dissect human bodies, Galen learned about anatomy from gladiators' wounds and by dissecting apes, sheep, pigs, goats and even elephants. In this way, he learned about the nervous system, and invented a complete system of treatment that remained the standard until just a few centuries ago. Not overtaxed with modesty, he was aware of his huge influence, writing: 'I have done as much for medicine as Trajan did for the Roman empire when he built roads and bridges. It is I, and I alone, who have revealed the true path of medicine. It must be admitted that Hippocrates already staked out this path ... but I have made it passable.'

Yet although Galen did study anatomy, the fact is that his anatomical knowledge came as much from animals as humans, and he did make basic errors. Such was his status, though, that a thousand years later many physicians – including those from the Islamic era – would insist that if Galen said it, it must be true, even if it contradicted the evidence of their own eyes.

Updating Galen

Not all of the Islamic medical scientists, though, were quite so reverential. When early-9th-century scholars like Hunayn ibn-Ishaq made the first translations of Galen into Arabic, these quickly became authoritative works on the subject. Yet despite Galen's status, ibn-Ishaq's readers began to wonder if Galen was always right about everything. Maybe even the great Greek medic had made mistakes.

We have already seen how Hunayn made a few minor updates to Galen's version of the anatomy of the eye. The first big challenge to Galen, however, came from a Persian medic called al-Razi (Rhazes in Latin) about half a century later.

Al-Razi, by all accounts, was never one to do things the conventional way. Born in the city of Rayy in 865, he apparently started out as a lute player. Very soon, though, he took up alchemy until, according to some sources, an experiment that went wrong damaged his eyesight. The story continues that after going for medical treatment for his eyes, he decided to take up medicine himself. It may be that he thought he could do a better job than the doctors who were treating him. If so, he was proved right. After excelling in training in Baghdad, he returned to Rayy as director of the city's hospital, where his lectures attracted many students and even patients. For a while, he was also director of the main hospital in Baghdad.

Al-Razi's dilemma

Something of al-Razi's character is revealed in the titles of books he wrote such as *On The Fact That Even The Most Skillful of Physicians Cannot Heal All Diseases* and *Why People Prefer Quacks and Charlatans to Skilled Physicians*. Like most major scholars of the age, he was a polymath, writing books on a range of disciplines from astronomy to natural history, but it was his medical work that sealed his place in history.

Al-Razi's innovations are a roll-call of the breadth of medical practice. He identified and described smallpox and measles, and his book on these two diseases was influential up until the 19th century. He also wrote one of the most important and comprehensive books about clinical practice which was called, in typically straight-forward fashion, *al-Hawy*, which means Comprehensive Book. It ran to 23 volumes and was an encyclopedia of Greek, Syriac, Persian, Indian and even Chinese medicine. And he discovered that fever is part of the body's defences.

Just as he had challenged his tutors as a student, al-Razi was also prepared to challenge the great Galen, writing, in his book *Doubts about Galen*, 'It grieves me to oppose and criticise the man, Galen, from whose sea of knowledge I have drawn much indeed ... Although this reverence and appreciation will and should not prevent me from doubting, as I did, what is erroneous in his theories.'

Historians argue over whether al-Razi was criticising just details in Galen, or whether he was concerned about Galen's entire system. At the heart of Galen's system was the idea, dating back to the time of Hippocrates in the 4th century, that good health required a balance between four kinds of body fluid, called humours. The humours were blood, phlegm, yellow bile and black bile. Each corresponded to one of the four elements of which the Greeks believed that everything is made – air, water, fire and earth – and with one of four qualities: warm and moist, cold and moist, warm and dry, and cold and dry. They also corresponded to four natural temperaments that people could have: sanguine, phlegmatic, choleric and melancholic.

The idea was that people fell ill when the four fluids somehow got out of balance. So the way to treat illnesses, logically enough, was to restore balance using a combination of diet and herbal drugs, as well as various invasive methods. For example, when someone went to the doctor complaining of fever or headache, if it was thought to be due to an excess of blood, then it was treated by letting some of the excess out of the body – simply by either cutting an artery and collecting the blood in a bowl, or pressing a small cup over a cut and letting the blood flow out into it. Bloodletting, or phlebotomy, remained a common treatment worldwide until the 1800s. Millions of patients through the ages, including George Washington, went through the ordeal of bleeding. Many actually died from loss of blood, but the treatment persisted. Bloodletting is still used by some cultures today.

Galen on trial

Al-Razi decided to conduct a trial to see if bloodletting worked as a treatment for meningitis. Two things are interesting about this trial. The first is the fact that he was not prepared just to accept Galen's idea as it stood, but wanted to put it to a proper test. The second thing is the methodology he used, which gives us an insight into his thinking. In his hospital he let one group of meningitis patients go untreated; but he treated another group by bloodletting in the normal way. Interestingly, the results of the trial supported Galen's view that bloodletting was an effective treatment – although few would accept that particular finding today.

In *Doubts about Galen*, al-Razi also seems to question the theory behind Galen's basic system. He asks if it is really true that giving a patient a hot drink would raise their body temperature even higher than that of the drink, as the theory of humours would seem to imply. It takes only a simple test, of course, to show that this is not true. If it is not, al-Razi suggests, there must be other control mechanisms in the body that the humours do not explain, but it is unclear how far he went with such ideas.

No-one really followed up al-Razi's doubts about the entire system of humours, though, and it was another thousand years before it was seriously challenged. However, in the South Asian *Unani* school of herbal medicine, it is still used as a basis of medical treatment by a majority of people in countries such as

Bangladesh, India and Pakistan. This is partly because modern healthcare is still unaffordable in these regions.

Ibn-Sina

Nevertheless, a few Islamic physicians gradually began to chip away at the edifice of Greek medicine, even as many more used it with dedication and, as Peter Pormann of Warwick University suggests, with some success. However, life for physicians was never quite so encouraging or so supportive as it had been in the early centuries of the Abbasid caliphates. By the time the next great figure in Islamic medicine, ibn-Sina (Avicenna), was born in 980, the empire was no longer under the control of a single caliph. The result was that ibn-Sina spent much of his colourful, varied life moving around trying to find a medical position that would pay him decently and give him the time to carry on with his other scholarly work.

Born near Bukhara in present-day Uzbekistan, ibn-Sina was something of a prodigy. By the age of ten, he knew not just the Qur'an but much Arabic poetry by heart, and by the age of sixteen had become a physician. Ibn-Sina proved his competence early on when he successfully treated the Samanid ruler of the eastern Islamic caliphate for a potentially life-threatening diarrhoeal infection. As reward, he was given access to the royal library at Bukhara, and certainly took advantage of it. His skill as a physician became almost legendary, even though the turbulent politics of the time kept him permanently unsettled, either stuck as a teacher or obliged to put himself at the whims of some prince or caliph.

Ibn-Sina managed to become one of the most famous philosophers, mathematicians and astronomers of his time, and wrote books on a range of scientific topics, a vast encyclopedia (one of the first ever written) and even poetry, including, perhaps, this verse in Omar Khayyam's *Rubaiyat* which is attributed to him, with its clear astrological symbolism and reference to his life's work:

> Up from Earth's Centre through the Seventh Gate
>   I rose
> And on the Throne of Saturn sate
> And many a knot unravelled by the Road
> But not the master-knot of Human Fate.

The Canon of Medicine

Ibn-Sina made a number of key astronomical observations, devised a scale to help make readings more precise, and made a string of contributions to physics, such as identifying different forms of energy – heat, light and mechanical – and the idea of force. He also noted that if light consists of a stream of particles, then its speed will be finite. The mathematical technique of 'casting out of nines', used to verify squares and cubes, is also attributed to ibn-Sina, among others. And he is believed to have suggested the fundamental geological idea of superposition – the concept that in rock layers, the youngest layers are highest – that would not be properly formulated until the 17th century.

Yet his fame, above all, is based on his book *al-Qann fi al-Tibb* (*The Canon of Medicine*). Consisting of some half a million words, this multi-volume book surveyed medical knowledge from ancient times to the present day. Its comprehensive, systematic approach meant that it became *the* reference for Arabic- and Persian-speaking doctors, and once it was translated into Latin it became one of the standard textbooks in Europe for six centuries, with some 60 editions being published between 1500 and 1674, according to the historian Nancy Siraisi.

Besides bringing together existing knowledge, the *Canon* contained many of ibn-Sina's own insights. He recognised, for instance, that tuberculosis is contagious; that diseases can spread through soil and water; and that a person's emotions can affect the state of their physical health. He also realised that nerves transmit pain and signals for muscle contraction. The *Canon* also contained a description of 760 drugs, and so became an important medical manual too.

Unlike al-Razi, though, ibn-Sina does not seem to have questioned the basic idea of humours. Indeed, it is possible that the clarity and care that was displayed in his work helped the concept to survive for longer than it might otherwise have done. But what he did do was give medicine a stronger evidence base for learning and moving forward, which has been the hallmark of the best medical practice ever since. Moreover, his *Canon* clearly laid out principles and procedures for testing new drugs.

There is no doubt that ibn-Sina was a proud, perhaps even arrogant and difficult man. Unfortunately, his absolute certainty that he was right (as he often was), along with his tendency to dismiss his critics as idiots, offended many, including his political patrons. This quality caused him to make some rather bold claims regarding the relationship of science to religion, and it meant that he would one day be charged with heresy.

Nature's laws

Like the scientist that he was, ibn-Sina firmly believed that there are laws of nature which cannot be violated. He believed that all physical phenomena have a known cause – an idea which also characterised his approach to medicine. This meant that he found it hard to envisage supernatural events such as healing miracles and bodily resurrection. For the mass of believers, miracles are an example of an active God bending the rules in order to prove the truths of religion to sceptics. But ibn-Sina believed that this does not happen. Early Islam did not seem to need miracles, and there is no record of the Prophet Muhammad having performed them. But by the 11th century, miracles were firmly established in Islamic theology as a route to gaining converts and supporters.

Ibn-Sina believed that there exists a single set of principles that can explain the nature of the physical universe, the reason for its creation, and the relationship between mind and body, and he made it his life's work to find connections between these apparently different fields, and ultimately to discover a theory of everything. This was an ambitious scheme, but then ibn-Sina, according to Yahya Michot of the Hartford Seminary in Connecticut, was always supremely confident of his abilities, and believed that God had deliberately made him brighter than the average individual.

So, according to ibn-Sina, miracles must have a physical explanation. To take one example: most Muslims believe that the world will end one day and that when this happens, every member of the human race will return from the dead in a physical form, ready to be judged by God for their conduct during their lifetime. But ibn-Sina held that such bodily resurrection defies the laws of nature, and he thought that the day of judgement might take a different form to that traditionally taught in religion. He also doubted the traditional view of heaven and hell, in part because of his belief that matter cannot be everlasting – no fire can burn forever. And he thought that heaven and hell might take the form of a state of mind, instead of a physical space. The example he gave to support his theory was that of pain. He postulated that if it is possible to feel pain without experiencing pain in the physical sense – such as during a bad dream – it ought to be similarly possible to experience heaven or hell without physically travelling to a different place.

The surgeon of al-Andalus

Although many other Islamic physicians made their contributions to the progress of medicine, two other names that stand out are Abul-Qasim al-Zahrawi (Albucasis) and ibn al-Nafis. Although al-Zahrawi was born half a century earlier than ibn-Sina, their lives overlapped for 33 years until al-Zahrawi died in 1013. But they lived at different ends of the Islamic empire, thousands of miles apart – ibn-Sina mostly in Central Asia and Persia, and al-Zahwari far away in al-Andalus.

Al-Zahrawi was the greatest surgeon of Islamic times, and devoted his whole life to surgery. When the palace at Medinat al-Zahra near Cordoba was sacked in 1010, the great royal library was destroyed. So we know little of his life, and it may be that much of his work was lost, too. One book that we do know a lot about is called *Kitab al-Tasrif li-man 'ajiza 'an al ta'lif*. This can be translated as 'Arrangement of Medical Knowledge for Someone Who Can't Compile a Manual for Himself'. Fortunately, it is usually just known as *Tasrif* ('Medical Knowledge'). It is essentially a practical manual comprising 30 volumes. The first is on general principles, the second is on diseases, symptoms and treatments, and volumes 3–29 are on pharmacology. The volume that has attracted most attention from historians, however, is volume 30, devoted to surgery.

Al-Zahrawi's surgery

Volume 30 was translated into Latin by Gerard of Cremona (who was also ibn-Sina's translator) in the 12th century, and it had a large influence on Western European surgery. Moreover, scholars are now discovering that it contains descriptions of relatively modern clinical techniques, such as 'Kocher's method' for treating a dislocated shoulder and the 'Walcher position' for simplifying difficult labours.

This volume also properly illustrated a range of surgical instruments. Some of the instruments were already in use. Others were devices that he developed or perfected. These included a variety of forceps to aid childbirth, an ingenious scissor-like instrument for extracting tonsils without choking the patient, a concealed knife for cutting abscesses without alarming the patient, and all kinds of useful hooks and pincers.

Another key innovation attributed to him is the use of catgut for sutures in internal operations. Catgut has the remarkable property of not arousing any immune reaction in the body, yet staying strong until it dissolves naturally after a few weeks. That makes it perfect for sutures inside the body, allowing surgeons to make internal stitches, then close the opening knowing that the catgut stitches will dissolve once the wound is healed, so the body does not have to be reopened. The use of catgut is mentioned in *Tasrif*, and it has proved invaluable to surgeons ever since.

Doctor of the heart

Ibn al-Nafis was born in Damascus in 1213 but subsequently moved to Cairo, which by then had some of the most advanced hospitals in the Islamic world, including the al-Mansuri where ibn al-Nafis became head physician. He wrote a medical book which is believed to have made him very rich, and which replaced ibn-Sina's *Canon* as the standard medical text in the Islamic world, though it had less impact in Europe. More significantly, he wrote commentaries on the work of Galen and ibn-Sina, correcting what he saw as some of their mistakes, for example on the pulse. But his real fame among historians in the West stems from a discovery made in 1924 that had some scholars rewriting medical history.

In 1924, a manuscript from ibn al-Nafis' *Commentary on Anatomy in Avicenna's Canon*, dating from 1242, was discovered in the Prussian State Library in Berlin. Galen (and later ibn-Sina) believed that blood seeped through from one side of the heart to the other through little holes in the septum that divided the sides. On examining many hearts, both alone and with witnesses, ibn al-Nafis could find no signs of such holes at all. He asserted instead that blood in the right ventricle of the heart must reach the left ventricle through the lungs alone, and not through small passages as Galen had maintained. Ibn al-Nafis had discovered what today we call pulmonary transit, or the lesser circulation.

Some writers and historians believe that ibn al-Nafis had in fact discovered the circulation of the blood.

Others, such as medical historians Emilie Savage-Smith of Oxford University and Peter Pormann of Warwick University, say that discovering pulmonary transit is not quite the same thing as demonstrating the continuous circulation of blood, which William Harvey did in 1628. This could be because ibn al-Nafis's description was uni-directional – it did not include the notion that blood returns from the left venticle to the right ventricle.

Ibn al-Nafis was, in many ways, part of the last generation of groundbreaking medical scientists in medieval Islam.

Ibn-Sina's critics

Ibn-Sina and other doctors of the period were very much experimentalists. They would play with different treatments and if something didn't work, they were happy to discard it and try an alternative method.

But theirs was not the only medical system in use at the time. Shortly after the time of ibn-Sina, Sufism began to take root, and Sufi ideas became more popular in Islamic territories. Sufism – in all its forms – has much to say on health and wellbeing, as well as the protection of the environment. One of its ideas is that asceticism (the suppression of worldly desires, shunning wealth and living on modest means) is both a route to good health and a way of pleasing God.

The proponents of such an approach to wellbeing included Abu Hamid al-Ghazali, an influential Sufi scholar and theologian from the 12th century. However, al-Ghazali was also critical of ibn-Sina and wrote a book called *The Incoherence of the Philosophers*. In this book, he took issue with what he considered to be the hubris of science and philosophy in their claims to be able to explain the natural world, which in al-Ghazali's opinion denied a space for God.

In one of his writings criticising ibn-Sina's attempts to produce a unified theory of the mind-body relationship, al-Ghazali wrote:

> Yet these philosophers ... they take the view that the equal balance of the temperament has great influence in constituting the powers of animals. They hold that even intellectual power in man is dependent on the temperament; so that as temperament is corrupted, intellect is also corrupted and ceases to exist. Further, when a thing ceases to exist, it is unthinkable in their opinion that the non-existent should return to existence. Thus it is their view that the soul dies and does not return to life – and they deny the future life – heaven, hell, resurrection and judgement. There does not remain, they hold, any reward for obedience or any punishment for sin.
>
> (*The Faith and Practice of al-Ghazali*, Oneworld, 2000)

The Prophet's Medicine

At the same time, yet another medical tradition, known as the Prophet's Medicine, also began to emerge, and has since grown into a global industry. In today's Muslim world, the Prophet's Medicine is every bit as popular as the herbal medicine of ibn-Sina, so much so that in the countries of south Asia (as well as among Muslims in countries such as Britain), the two systems have either merged, or are regarded by patients as being one and the same.

We know a good deal about the Prophet's Medicine, thanks to a 14th-century manual that is still in print to this day. It is called *Medicine of the Prophet*, and its author was a scholar of religion from Damascus named ibn-Qayyim al-Jawziyya. He belonged to the same Islamic tradition as Ahmad ibn-Hanbal, the Baghdad jurist from the 9th century who was tortured by his caliph for refusing to sign up to state-sponsored rationalism. This is known as the Hanbali school, which is still popular today in Saudi Arabia and forms the basis of the country's culture and laws.

The Hanbali school took a dim view of both experimental medicine and the Sufism of al-Ghazali. The arguments against Sufism were powerfully made: for example, ibn-Qayyim could not understand how God would look kindly on a Muslim who intentionally makes himself and his family destitute and thus dependent on others for charity. If anything, ibn-Qayyim believed that such extreme poverty was more likely to lead people away from religion than to make them better Muslims. Similarly, ibn-Qayyim also saw experimental medicine as an inferior system because it was often uncertain and open to being improved or superseded by the next discovery.

The solution, in his eyes, was a system of medicine based on medical references found in the Qur'an itself, and many more that appeared in biographical works on the life of Muhammad. Of divine origin, these would be less open to challenge and change than any alternative. At the core of the Prophet's Medicine is the idea that the Qur'an, being God's word, can also be regarded as the final word in healthcare and in healing.

The Prophet's Medicine, however, also has a strong folk-healing dimension to it. For example, it includes the idea that good (or bad) health is linked to the motion of the planets and can therefore be affected by astrology; and that evil spirits, or black magic, also have the power to make people fall ill. Also in this tradition is the idea that deviation from religion is a potential cause of disease: in other words that illness might be a test, or a punishment from God for poor behaviour. In such cases, a doctor's prescription might also include instructions to perform special prayers, fasting or charitable giving.

Plagued by controversies

One of the best-known examples from history of the differences between the experimental approach to medicine and that of the Prophet's Medicine is in the treatment of plague. According to ibn-Qayyim, the Prophet Muhammad is believed to have said the following about plague: 'Plague is punishment sent on those who disobey God. Do not enter a land where you hear of plague. And do not flee if it breaks out in your own land.' In addition, he is also reported to have said: 'Plague is martyrdom for every Muslim.'

The authenticity of these reports is open to question. However, for ibn-Qayyim, the message is clear: if you get the plague, you need to stay where you are. Similarly, 'plague is martyrdom' could be taken to mean that there is little or no point in trying to treat the disease. Referring to experimentalists such as ibn-Sina, ibn-Qayyim said in his book: 'The physicians have nothing whereby to repel these illnesses and their causes, any more than they have anything to explain them.' Later, he added: 'Have complete trust and confidence in God and to endure patience and accept with contentment His decrees.'

In spite of their radically different approaches to healthcare, the Prophet's Medicine and experimental medicine eventually began to merge in the Islamic world. This development holds important lessons for how new knowledge is absorbed in Islamic countries in the present day.

9

Astronomy:
The Structured Heaven

Have they not looked at the heaven above them – how We structured it and adorned it and how it has no rifts?

The Qur'an

In few places is the night sky such a jewelbox of stars as above Arabia, the birthplace of Islam. The desert air often ensures good visibility, and for trading caravans travelling at night to avoid the heat of the day, the stars must have been familiar guides in a featureless landscape long before the coming of the Prophet Muhammad. It may be, then, that the Arabic names of so many of our stars – Aldebaran, Rigel, Formalhaut, Betelgeuse, Deneb, Altair and many more – come from these ancient days. The coming of Islam gave astronomy an exalted status and ensured that these names have survived to become the names astronomers use even today.

There were many reasons for the prominence of astronomy in Islam, besides natural curiosity and the desire for knowledge. Travel across a vast empire, encompassing large areas of sea and desert, called for navigational aids which only the stars could provide. And astrology – which in older Arabic was the same word as astronomy – still attracted caliphs with its power to predict, despite the objections of many theologians. The rise of the Abbasids, for instance, brought some of the ancient Persian Zoroastrian tradition of astrology right into the heart of Islam, and each Abbasid caliph had his own personal astrologer from the Naubakht dynasty. Many patrons also called on astronomers to give them astrological 'inside information', and many historians say that the pursuit of astrology was a key reason for the development of observatories.

Astronomical demands

At least three Islamic teachings had implications for astronomy. First of all, Muslims were told to pray daily at five specific times – sunset, late evening, dawn, soon after midday and late afternoon. In the days before alarm clocks, this was by no means as easy as it sounds. The only way to be sure when to pray would be to observe the angle above the horizon of the sun or the stars in the sky. And if it was important to do it at precisely the correct time, then the more accurately that could be calculated, the better. It took a concerted effort by astronomers to work out how to do this in such a way that the appointed time had not passed before one had calculated it.

The mathematical method of telling the time at night, for instance, is to work out the unknown sides or angles of a vast triangle between the earth and the sky, from the known sides and angles. At one corner of the triangle is a particular star. At another is the north celestial pole – the point in the sky about which all the stars rotate. The third is the zenith, the highest point to which the star rises during the night. Working this out pushed astronomical calculation and the related mathematics of trigonometry to new levels. It also helped to drive innovations in the astrolabe, a device for computing angles developed in Greece. After suitable modifications, telling the time on a starry night eventually became easier.

Standing towards Mecca

Second, Muslims are expected to pray towards the Ka'bah in Mecca. This direction is called the *qibla*, and many astronomers and mathematicians worked hard at getting the *qibla* right. This is a surprisingly tricky problem because, as the surface of the earth is curved, it is necessary to work out a particular direction across a curved surface. (Interestingly, there was never any question in the Islamic world at this time about the earth not being round.) This is complex spherical geometry, and also demands very accurate observations of the reference points in the night sky – since even the tiniest error could throw the calculations out.

Third, the Islamic calendar consists of twelve lunar months in a year. Each lunar month starts with the moon's first visible crescent. Predicting exactly when that crescent will appear presented early Muslim astronomers with a real challenge. The peoples of pre-Islamic Arabia used a mixed lunar/solar calendar, in which eleven days would be added to the 354 days of a lunar year to bring it in line with a solar year. This process is called intercalation; however, it is forbidden in the Qur'an, which is why Islam uses a lunar calendar to this day.

As the ability to tell the time accurately was useful in so many ways, most mosques employed an official time-keeper or *muwaqqit* to help the faithful pray as directed, and on time. *Muwaqqit*s were competent astronomers, and so thousands of them across the empire were keeping records of planetary motion, which they added to the growing body of increasingly accurate tables that were produced throughout the Islamic era. Some historians, such as David King, have begun to explore religious astronomy in medieval Islam, and realised that there was a whole other dimension, barely yet appreciated, concerned with matters such as the alignment of mosques and other buildings.

Clearly, the early Arabic-speaking Muslims had a long heritage of their own astronomical observations to help them, but from the beginning of the Abbasid era they also worked closely with astrologers and astronomers from Persia. When astrologers like the Naubakhts and the Persian Jewish Masha'allah ibn-Athar (Messahala) came to Baghdad, they brought with them tables called *zij*, which showed the positions of the sun, moon, planets and stars, compiled over many centuries.

Faith in the stars

At the same time, the Islamic empire's rulers were happy to pay for the necessary infrastructure of astronomy, such as observatories and astronomical instruments. They took a strong interest in the activities of the institutions they funded, and got to know the leading astronomers of the day on a personal level. A few of them were also practising astronomers.

Such interest helped to ensure that one of Islam's seminal contributions to modern science has been in the patronage and development of observatories. The first of these were built in 9th-century Baghdad and 10th-century Cairo, though the latter remained incomplete. In later centuries, these would be eclipsed by bigger and better ones in Istanbul, Maragha, and Samarkand in modern-day Uzbekistan. In the majority of these examples – and in many more – the observatories would contain specialised instruments such as quadrants, armillary spheres and astrolabes.

Islam's observatories were usually paid for from the private funds of rulers such as al-Mamun in Baghdad and al-Hakim in Cairo. In addition, the relationship between a ruler and his chief astronomer was often close – rather like a head of government and his or her chief scientist today. For example, the scientist Hassan ibn al-Haitham worked in Fatimid Cairo during the 11th century under the ruler al-Hakim; ibn al-Shatir worked as an astronomer and timekeeper in Damascus' largest mosque during the 14th century; ibn-Sina worked in 11th-century Central Asia under a number of different rulers; and Nasir al-Din al-Tusi, who directed the Maragha observatory, worked directly for the ruler Helagu Khan – he is believed to have accompanied Helagu during the latter's assault and sacking of Baghdad. Finally, one of the largest observatories was the Samarkand, established in the 15th century by the governor Ulugh Beg, who was a keen amateur scientist.

We know that observatories were popular with many more of Islam's caliphs and rulers. We also know that most observatories did not qualify for financing under Islam's special system of religious endowments, which helped to pay for mosques, schools, colleges and hospitals. As a result, the vast majority of observatories did not survive beyond the lifetime of their founding patron. Whereas mosques, hospitals, universities and schools built during the Islamic era survived many centuries, the longest-lasting observatory had a working life of just 30 years. In almost every case, once a patron died, it wouldn't be long before his observatory would follow him to the grave.

Such a relatively short life also suggests that, while observatories were undoubtedly important in the conduct of religion, they were not seen as essential or critical for the conduct of faith – at least not in the same way that a mosque or a hospital might have been. Another reason for this is their use in astrology.

Historians of Islamic astronomy, such as the late Aydin Sayili from Turkey and David King, agree that the needs of faith did help to drive astronomy. At the same time, there is no doubt that, as far as rulers, governors and caliphs were concerned, astrology was an important motivation for their funding of, and interest in, the work of astronomers. In some ways, this interest in astrology was similar to their desire to fund the translation of Greek works of astrology into Arabic.

In contrast to its relationship with modern science today, astrology during the Islamic middle ages (as in much of the Latin West) was regarded by the ruling classes as an important application of astronomy, or as applied astronomy. The argument went that if the moon could influence tides, then surely it should be possible for planets to influence other physical things, as well as events and people.

A few observatories were based inside or close to royal palaces, and most of the leading astronomers were often asked for astrology-based advice by their rulers. This included advice on political appointments, impending wars and invasions, as well as advice on who (and when) to marry. This meant that astronomers and astrologers were very powerful people, and had the ability to influence royal decisions.

The astrolabe

One of the early astronomers was Ibrahim al-Fazari. He was either a Persian or, judging from his name, an Arab who learned his craft in Persia. He clearly knew his astronomy, for under the direction of the Caliph al-Mansur (754–75 CE) he made a highly technical translation of the ancient Indian astronomical text by Brahmagupta which came to be known as the *Sindhind*. The Muslim willingness to learn from other astronomical traditions is part of the reason for their extraordinary success. But the translation of the *Sindhind* was valuable not only for its astronomical insights – it is thought that it also brought Indian numerals into the Arabic world for the first time; a task later completed by al-Kharwizmi, who also produced an improved version of the *Sindhind*.

Under the instructions of Caliph Harun al-Rashid, al-Fazari also made the first known astrolabe in the Islamic world. In the hands of the Arabic craftsmen and astronomers, the astrolabe became one of the most beautiful scientific instruments ever made. It was not only the wonderful craftsmanship that made these brass mechanical computers such attractive objects; it was the increasingly intricate and precise design that meant that they were the medieval equivalent of a GPS unit. The astrolabe was a model of the universe that you could hold in your hand. By using it to measure the angle of stars and the sun above the horizon, it could tell you anything from your current latitude to where stars will appear in the sky. It became the main navigational aid for many centuries, celebrated in Chaucer's *Treatise on the Astrolabe*, until it was superseded by the simpler quadrant.

The coming of Ptolemy

It was just a few years after al-Fazari's astrolabe, in the reign of Caliph al-Mamun, that astronomy really began to take off. The catalyst was the translation of a number of the works of the Greco-Roman astronomer Ptolemy. A Syriac version of Ptolemy's *al-Majisti* or *Almagest* was followed by three Arabic versions, Arabic versions of his *Planetary Hypotheses* (which explained his theory of how the planets moved) and his *Handy Tables* for predicting the movement of the planets and stars. The impact of Ptolemy's books was dramatic, and would shape the course of Islamic astronomy throughout the medieval period.

First of all, though, the demand was for up-to-date and accurate *zij*, the tables of celestial movements. New tables were needed for religious purposes, and as navigational aids. And so began a massive and never-ending project to produce *zij* based on both observations and recalculations. The astronomers who produced these tables could be found at all levels of society. They were employed by patrons, they worked in mosques, and many were enthusiastic amateurs.

The great skywatch

To get the new observations, rulers and wealthy patrons began to establish observatories. The first were set up in the 820s by al-Mamun in Baghdad, and on Mount Qasiyun near Damascus. Their task was to reconcile the data from the three different traditions – Persian, Indian and Greek. Thereafter, all new *zij* were based essentially on the Ptolemaic model of the *Handy Tables*. Other famous observatories were at Rayy (near modern Tehran), Isfahan and Shiraz. Over the centuries, observatories became ever bigger and more spectacular; partly no doubt for status reasons, but also to achieve greater and greater precision, with giant sextants and quadrants as big as artificial ski-slopes. The largest and most spectacular were at Maragha in Persia and at Samarkand – these obervatories were set up in the 13th and 15th centuries by invading Mongol and Turkic descendants of Genghis Khan and Tamerlane whose hordes invaded the eastern Islamic empire and took over its institutions. The Samarkand observatory, run personally by Tamerlane's grandson Ulugh Beg, was the largest of all, and its great 130-foot radius arc can still be seen today plunging into the ground.

Using observatories like these, along with increasingly sophisticated calculations in spherical geometry and trigonometry, the Arabic astronomers gradually made more and more accurate measurements of the earth and the heavens. They calculated the tilt of the earth on its axis, reaching a figure that was remarkably close to the modern one, and refined the measurement of precession – the slow rotation of the earth's tilt over nearly 26,000 years. They also calculated the circumference of the earth to be 24,835 miles (compared with current measurements of 24,906 miles), and measured how the earth's furthest point from the sun shifted by a few seconds each year.

Islamic superstars

One observation in particular stands out. In 1006, a brilliant new star suddenly appeared in the night sky. A young astronomer in Cairo called ibn-Ridwan described this startling event with the precision that became the hallmark of the Arab astronomers:

> The sun on that day was 15 degrees in Taurus and the spectacle in the 15th degree of Scorpio. The spectacle was a large circular body, two and a half to three times as large as Venus. The sky was shining because of its light. The intensity of its light was a little more than a quarter that of moonlight. It remained where it was and it moved daily with its zodiacal sign until the sun was in sextile with it in Virgo, when it disappeared at once.

So exact and full was this young boy's description that astronomers can be certain today that what he was seeing was a supernova 7,000 light years from earth which they have named Supernova 1006, after the year when it was first seen.

Nearly all the great Islamic scholars contributed their ideas and observations to astronomy, from al-Khwarizmi and ibn-Sina (Avicenna), to ibn-Rushd (Averroes) and Musa bin Maymun (Maimonides). The *zij* tables of al-Khwarizmi and al-Battani were picked up in Spain by astronomers like Maslama al-Majriti in the 10th century, who not only updated them with his own remarkable observations but translated them into Latin, and so began the gradual process of the transmission of Islamic astronomical data and ideas into Europe.

Over the centuries, hundreds of *zij* were produced by Islamic scientists. On the whole, the new observations and more precise calculations meant that they became more accurate as time went on. And yet it was not simply a matter of getting better and better observations and calculations. Each time a new table was issued it was accurate for a little while, but sooner or later a discrepancy began to appear between the predicted position of the planets and their actual position. There was clearly a flaw in Ptolemy's basic model, and as the centuries went on, this problem began to occupy Islamic astronomers more and more.

The Ptolemaic system

Very little is know about Claudius Ptolemy, beyond the fact that he was Greek and lived in Alexandria between 90 and 168 CE. Yet he wrote two profoundly influential works. One was his *Geography*, which became the standard atlas of the world for the next 1,300 years. The other was the *Almagest*. This work provided a complete model for the movement of the sun, moon, planets and stars that was developed over more than five centuries, but came to be called the Ptolemaic system. It was an entirely mechanical model, based on the scientific world view of ancient Greek thinkers such as Aristotle. It is all based on the rotation of perfect spheres, since nobody could contemplate any other shape for heavenly bodies until Kepler introduced the idea of less-than-perfect spheres in the early 17th century.

At the centre of Ptolemy's system is the fixed earth. Around it rotates a vast sphere carrying with it, in a series of seven perfectly spherical layers, the sun and the moon, the five planets that were known at the time and, furthest out, the stars. As these transparent 'crystal' spheres rotate, they carry all the celestial bodies with them, so we see them moving through the skies. This was not simply meant to be an attractive theoretical picture of how things were, but a model for predicting the movements of these bodies precisely – and this is where it became complicated.

Matching theory with reality

With the earth fixed in one place, it's difficult to get the observed movements of the heavenly bodies to match the model – particularly the movements of the planets. Unfortunately, only the stars move in perfect circles. It is bad enough that the sun's path through the sky changes through the year. How could it possibly do that if it were simply swept around on the surface of a ball? The planets make even less sense, since their path seems even more variable than that of the sun. That's how they got their name, which is the Greek word for 'wanderers.' In particular, the planets not only seem to shift a little further eastwards against the background of the stars each night; they also seem to loop back and travel westwards for a few months each year, a phenomenon called 'retrograde motion'. Nowadays this looping back is easily explained by the fact that the earth is constantly overtaking slower-moving planets further out from the sun, and constantly being overtaken by faster-moving planets closer in. Yet if the earth is fixed, this motion is very hard to explain, and it caused astronomers problems for millennia.

Because of the model that was accepted at the time, the ancient Greeks had to think in terms of perfect circles and perfectly uniform motion. And yet they had to get their model of the spheres to match the observed movement of the planets, sun and moon precisely, or it would not work for prediction. Over the centuries, they gradually found answers to all these problems, or at least they seemed to. In the 3rd century BCE, Apollonius suggested that there are wheels within wheels. All the while the planets are going round in a big circle (the 'deferent'), he suggested, they are also whirring round in a small circle or 'epicycle' too, like some celestial waltzer. A century later, Hipparchus then 'explained' the sun's motion by suggesting that its rotation is eccentric – that is, its centre of rotation is slightly offset from the centre of the earth.

The problem is that these theoretical motions still didn't match real-life observations. So with mind-boggling ingenuity, Ptolemy combined epicycles with doubly offset eccentric rotations for the planets around points called 'equants' to create a clockwork machine of immense complexity. Remarkably, though, it seemed to work, and its predictions were always pretty nearly right, which is the main reason why the Islamic astronomers took it on board so completely. But it was the 'pretty nearly' that eventually caused to them to begin asking questions. There was always that gradual slippage between the tables and the observations that meant that the tables constantly had to be updated.

Doubts about Ptolemy

Gradually, Islamic astronomers began to think that there might be problems with the Ptolemaic model. The model was meant to be a picture of the real world, describing how the celestial bodies actually moved. But the continual adjustments that it needed drew their attention to its basic conceptual flaws. Arab astronomers began to ask how some of Ptolemy's epicycles and equants could work in the real world. Just as al-Razi had written his *Doubts about Galen*, so the great polymath ibn al-Haitham (Alhazen) wrote his *Doubts about Ptolemy* (*al-Shukuk ala Batlamyus*). And just as al-Razi had simply raised questions, so did ibn al-Haitham. He focused on Ptolemy's concept of the eccentric motions and equants, because he could not see how they could possibly be real. Real objects, he knew, simply did not move like that. A real sphere simply can't rotate off-centre and yet stay in the same place. Yet 'no motion exists in this world in any perceptible fashion,' ibn al-Haitham argued, 'except the motion of [real] bodies.' There just had to be a central point about which everything rotated.

A few centuries later, in the 12th century, ibn-Rushd (Averroes) went further, declaring that:

> To assert the existence of an eccentric sphere or an epicyclic sphere is contrary to nature … The astronomy of our time offers no truth, but only agrees with the calculations and not with what exists.

And if the calculations were beginning to look uncertain, too, then it was clear that the smooth clockwork of the Ptolemaic system was begin to rattle alarmingly.

Tuning the model

Over the next few centuries, Islamic astronomers began to make adjustments to Ptolemy's model to try to make it conform with motion that was believable in the real world. Interestingly, no major astronomer ever really gave much time to the idea that the earth moved, even though it had been suggested – because it did not correspond to any real motion that they could imagine. The idea of the fixed earth at the centre of concentric spheres, on the other hand, certainly did.

So with as much ingenuity as the Greeks had shown, the Arab astronomers began to tinker and fiddle to get rid of equants and make all the celestial motions, as far as they could see, possible in reality. A key breakthrough was made by a brilliant astronomer born in the Persian Khorasan city of Tus in 1201. Nasir al-Din al-Tusi was born into a frightening time, when the forces of Genghis Khan were just beginning to spread across Asia.

By the time al-Tusi was thirteen years old the Mongols had dealt with China, and were soon storming west into Central Asia, generating horror story after horror story as they moved on towards the heartland of Islam. As they neared Tus, the young al-Tusi was sent away to Nishapur. Nishapur was not attacked at first, but he must have heard the devastating news that his home town had been ravaged by the Mongols. Nowhere, it must have seemed, was going to be safe, especially on the plains where the Mongol horsemen could ride easily. This may be why al-Tusi decided to take a job with the governor of Alamut, up in a secure mountain fortress. Alamut was the centre of power of the Ismaili branch of Islam, and al-Tusi made himself at home in the city, taking on the Ismaili faith.

The Khan's astronomer

For 30 years, Alamut was a place of safety, and there al-Tusi devoted himself to the study of astronomy and mathematics, writing a number of important books which only reached Europe much later, including his radical rethink of Ptolemy. Even Alamut, however, would not be safe forever. In 1256, the Mongols arrived on the plains below the fortress under the leadership of Genghis Khan's grandson Helagu, and soon managed to find a way into the impregnable fortress, perhaps by treachery.

Extraordinarily, al-Tusi not only survived the general slaughter but was taken on by Helagu as his personal astrologer. Not only that, but Helagu built for him the biggest, best-equipped observatory that had yet been constructed, at Maragha in Persia. It had the largest quadrant ever made, four metres across and made of solid copper, and a library that soon possessed some 400,000 books. Interestingly, the line of communication that opened up to China across the vast Mongol empire gave al-Tusi access to a whole new set of astronomical data and ideas, while Muslim astronomers who had been trained at Maragha trekked eastwards to create a new generation of observatories in China.

The Tusi Couple

Recalculations made at Maragha enabled al-Tusi to compile the most complete and accurate set of tables so far, known as the *Zij al-Ilkhani* after his patron. He also clearly established trigonometry as a separate branch of maths independent of spherical geometry, dramatically streamlining calculations about distances and directions in the heavens. But his greatest breakthrough was in finding a way to get rid of most of the equants from the Ptolemaic model and to replace them with believable uniform motion. He did this with an idea that came to be called the Tusi Couple.

The Tusi Couple was a way of showing how realistically uniform motion in circles can actually end up making something appear to move in a straight line. This sounds impossible, but it works like this: imagine a wheel rolling around the inside wheels of a drum. If the wheel is exactly half the diameter of the drum, then any point on the rim of the wheel will appear to move in a straight line across the drum.

Using this idea, al-Tusi was able to simplify the Ptolemaic system and get rid of the problematic equants for the sun and the 'upper' planets (Saturn, Jupiter and Mars). Yet he couldn't get rid of them for Mercury. The moon was even more of a problem. The Mercury problem was partly answered at the turn of the 14th century by al-Tusi's student and colleague Qutb al-Din al-Shirazi, by combining al-Tusi's ideas with those of another 13th-century Arab astronomer, Mu'ayyad al-Din al-'Urdi. Half a century later, ibn al-Shatir, who worked as a *muwaqqit* in the Great Mosque of Damascus, went further and found a way to get rid of all extra motions but the epicycles, including those for the moon.

So by the late 14th century, the Islamic astronomers had completely overhauled Ptolemy's system to produce a model that not only predicted the motions of the heavenly bodies with a high degree of accuracy, but also made sense in terms of the contemporary understanding of how the real world worked. This was an enormous achievement. The problem was that it was wrong, as we now know.

Moving the earth

With the benefit of hindsight, it is easy to see that the Islamic astronomers' basic assumptions were flawed. Of course Copernicus showed in the mid-16th century that the earth does move, circling around the sun with the planets. But even this concept failed to give correct predictions until Kepler showed that the paths of the planets through space are not perfectly circular, but slightly elliptical. And it would have made no sense in terms of existing theories of how the celestial machine held together. It required the addition of Newton's theory of gravity to complete the picture and show how it all worked.

In conventional accounts, the narrative seems to leap straight from Ptolemy to Copernicus, and to show how Copernicus had the great insight to see that the earth is not fixed, as Ptolemy said it was, but circles around the sun and spins on its axis. In this narrative, the ultimate Islamic contribution to the big picture seems comparatively small or even misguided. The Arab astronomers may have been diligent and ingenious, it seems, but they were barking up the wrong tree in backing the fixed-earth model, and it required Copernicus's brilliant insight to set things right.

Copernicus acknowledged that some of the data he needed to prove his theory came from the charts of al-Battani and al-Bitruji, but that was all that apparently came from the Arab astronomers. Yet there are clues that this is not the full story.

Islamic source

In 1957, the historian Otto Neugebauer noticed a similarity beween an illustration in Copernicus's first key book *Commentariolus* (1514), in which he first set out his idea that the earth moves, and one in ibn al-Shatir's book in which he answered the problems of the moon's motion. The similarity was so striking that it seemed hard not to believe that Copernicus had seen ibn al-Shatir's book. Intrigued, Neugebauer delved deeper for connections between Copernicus and the Islamic astronomers, and soon found another apparent illustration match in Copernicus, this time with al-Tusi's 1260 *Tadhkira*, in which he explains the Tusi Couple. Again the similarity was marked, even including an apparent mistake in the copying of an Arabic letter in al-Tusi's illustration.

Many historians now believe that Copernicus drew directly from the work of the Islamic astronomers in providing proofs for his theories. Recent research has suggested that West European astronomers were far more aware of Arabic work at the time than was imagined. Indeed many may actually have spoken, or at least read, Arabic, including Guillaume Postel, a lecturer at Paris University in the early 16th century, whose highly technical notes in Arabic can clearly be seen on an Arabic astronomical text in the Vatican library.

The Arab contribution

Of course, Copernicus made the great breakthrough suggestion that the earth moved, but the argument is that it was simply yet another step down the road away from the Ptolemaic model. Indeed, at the time, in some ways it seemed like a backward step, since ibn al-Shatir's work had matched a believably real theory with observations to a remarkable degree. Yet Copernicus's idea did not. No one at the time could explain how the universe could possibly fit together without the earth at its centre – and Copernicus's model made considerably less accurate predictions than ibn al-Shatir's. These problems, as much as any theological problems that the Roman Catholic Church might have had, needed to be solved before most astronomers could accept that the earth moves.

There is no doubt that Copernicus's idea of a heliocentric (sun-centred) universe was a seismic shift in scientific thinking. But it was a revolution waiting to happen. The way was paved by the gradual chipping away at the edifice of the Ptolemaic system over the centuries by countless Arabic astronomers, both with their observations and their often ingenious theories.

10

Number: The Living Universe of Islam

In Greek mathematics, the numbers could expand only by the laborious process of addition and multiplication. Khwarizmi's algebraic symbols for numbers contain within themselves the potentialities of the infinite. So we might say that the advance from arithmetic to algebra implies a step from being to 'becoming', from the Greek universe to the living universe of Islam.

George Sarton, *Introduction to the History of Science*, 1927

In many areas of science, the contribution of early Islam is sometimes open to interpretation and shifts of opinion, but when it comes to numbers and mathematics the legacy is immense and indisputable. The very numbers in use in our world every day for everything from buying food to calculating the spin on an atomic particle are called Arabic numerals, because they came to the West from scholars who wrote in Arabic. What's more, with al-Khwarizmi's algebra, these scholars provided us with the single most important mathematical tool ever devised, and one that underpins every facet of science, as well as more everyday processes.

Abu Ja'far Muhammad ibn-Musa al-Khwarizmi is the great hero of Arabic mathematics. Like so many of the early Islamic scholars, his interests were wide-ranging, but it is in the world of numbers that his legacy lies. Little is known about the man, and much of his reported life story could be speculative. It seems likely that he was born in what is now Uzbekistan south of the Aral Sea in Central Asia.

Some scholars say his father was a Zoroastrian, and that he was brought up in this faith which dates back to the time of ancient Sumeria. Others say that this is to completely misinterpret the records. All we do know is that al-Khwarizmi was born about 786, the year Harun al-Rashid came to power, and that when Harun's son al-Mamun set up the House of Wisdom, al-Khwarizmi was there studying. There is a story that he was summoned to al-Mamun's sickbed to make an astrological prediction about his health. Sensibly, al-Khwarizmi predicted that the caliph would live another 50 years. In fact, al-Mamun lived just ten more days. Al-Khwarizmi lived much longer. Other accounts say that he was actually one of al-Mamun's key advisors.

Numbers from India

One of al-Khwarizmi's greatest contributions was to provide a comprehensive guide to the numbering system which originated in India about 500 CE. It is this system, later called the Arabic system because it came to Europe from al-Khwarizmi, that became the basis for our modern numbers. It was first introduced to the Arabic-speaking world by al-Kindi, but it was al-Khwarizmi who brought it into the mainstream with his book on Indian numerals, in which he describes the system clearly.

The system, as explained by al-Khwarizimi, uses only ten digits from 0 to 9 to give every single number from zero up to the biggest number imaginable. The value given to each digit varies simply according to its position. So the 1 in the number '100' is 10 times the 1 in the number '10' and 100 times the 1 in the number '1'. An absolutely crucial element of this system was the concept of zero.

2. Numerals through the ages: Brahmi numerals from India in the 1st century CE, the medieval Arabic-Indic system, and the symbols used today.

This was a significant advance on previous numbering systems, which were often cumbersome with any large numbers. The Roman system, for instance, needs seven digits to give a number as small as, for example, 38: XXXVIII. Arabic numbering can give even very large numbers quite compactly. Seven digits in Arabic numerals can, of course, be anything up to 10 million. What's more, by standardising units, Arabic numerals made multiplication, division and every other form of mathematical calculation simpler.

This system quickly caught on, and has since spread around the world to become a truly global 'language'. Along with the numbers, English also gained another word, 'algorithm', for a logical step-by-step mathematical process, based on the spelling of al-Khwarizmi's name in the Latin title of his book, *Algoritmi de numero Indorum*. The new numbers took some time to embed themselves in the Islamic world, however, as many people continued with their highly effective and fast method of finger-reckoning.

The discovery of algebra

Al-Khwarizmi's other major contribution also introduced a new word to the language, 'algebra', and a whole new branch of mathematics. What is interesting is that in developing algebra, al-Khwarizmi had something eminently religious in mind, not just abstract theory. According to one report, he wrote his book on algebra in response to a request from the caliph to come up with a simple method for calculating Islamic rules on inheritance, legacies and so on. In his introduction to the book in which he describes algebra, he says that the aim is to work with 'what is easiest and most useful in mathematics, such as men constantly require in cases of inheritance, legacies, partition, lawsuits, and trade, and in all their dealings with one another, or when measuring lands, digging canals and making geometrical calculations'. Al-Khwarizmi would typically introduce a problem like this:

> Suppose that a man who is terminally ill allows two of his slaves to buy their freedom. The price of one slave is 300 dirhams. This slave dies, leaving a daughter and two sons. He also leaves property worth 400 dirhams. Then his former master dies, and he leaves three sons and three daughters. How much do each of the children receive in inheritance?

Although we now associate algebra entirely with the idea of symbols replacing unknown numbers in calculations, al-Khwarizmi did not actually use symbols, for he wrote everything out fully in words, and for the unknown quantity he would not use 'x' or 'y' but the word 'shay'. It was in his way of handling equations that he created algebra.

Completing and balancing

In his work on algebra, al-Khwarizmi worked with both what we now call linear equations – that is, equations that involve only units without any squared figures – and quadratic equations, which involve squares and square roots. His advance was to reduce every equation to its simplest possible form by a combination of two processes: *al-jabr* and *al-muqabala*.

Al-jabr means 'completion' or 'restoration' and involves simply taking away all negative terms. Using modern symbols, *al-jabr* means simplifying, for instance, $x^2 = 40x - 4x^2$ to just $5x^2 = 40x$. *Al-muqabala* means 'balancing', and involves reducing all the postive terms to their simplest form. *Al-muqabala* reduces, for instance, $50 + 3x + x^2 = 29 + 10x$ to just $21 + x^2 = 7x$.

In developing algebra, al-Khwarizmi built on the work of early mathematicians from India, such as Brahmagupta, and from the Greeks such as Euclid, but it was al-Khwarizmi who turned it into a simple, all-embracing system, which is why he is dubbed the 'father of algebra'. The very word algebra comes from the title of his book, *al-Kitab al-mukhtasar fi hisab al-jabr wa'l muqabala* or *The Compendious Book on Calculating by Completion and Balancing*.

Universal solutions

By completing and balancing, al-Khwarizmi could reduce every equation to six simple, standard forms and then show a method of solving each. He went on to provide geometrical proofs for each of his methods, which is where the debt to Euclid comes in. So what he was saying was that he could use his notation and the rules of *al-jabr* and *al-muqabala* to simplify *any* kind of problem, especially ones involving the tricky quadratic. Any problem – including things not yet thought of – could be reduced into one of his special six categories. It is for these reasons that later mathematicians such as Galileo and Fibonacci held him in such high regard.

Simplifying quantities into symbols (even quadratics) dates back to the time of mathematicians such as Diophantus and Pythagoras from Greece, as well as Brahmagupta in India. But Roshdi Rashed, a historian of mathematics at the National Scientific Research Centre in Paris, says that al-Khwarizmi's contribution represents a forward step for several reasons: although people were working on solutions to quadratics before him, al-Khwarizmi helped to find a suite of solutions that could solve all conceivable kinds of quadratics. No mathematician had done this before him.

Higher maths

Beyond al-Khwarizmi, many other Arabic-speaking scholars explored mathematics. Indeed, it was fundamental to so many things, from calculating tax and inheritance to working out the direction of Mecca, that it is hard to find a scholar who did not at some time or other work in mathematics. But it wasn't just practical applications that fascinated many of them, and they began to push mathematics to its limits.

In the early 11th century in Cairo, Hassan ibn al-Haitham, for instance, laid many of the foundations for integral calculus, which is used for calculating areas and volumes. Half a century later, the brilliant poet/mathematician Omar Khayyam found solutions to all thirteen possible kinds of cubic equations – that is, equations in which numbers are cubed. He regretted that his solutions could only be worked out geometrically rather than algebraically. 'We have tried to work these roots by algebra, but we have failed', he says ruefully. 'It may be, however, that men who come after us will succeed.'

The poetic mathematician

Omar Khayyam is one of the most extraordinary figures in Islamic science, and tales of his mathematical brilliance abound. In 1079, for instance, he calculated the length of the year to 365.24219858156 days. That means that he was out by less than the sixth decimal place – fractions of a second – from the figure we have today of 365.242190, derived with the aid of radio telescopes and atomic clocks. And in a highly theatrical demonstration involving candles and globes, he is said to have proved to an audience that included the Sufi theologian al-Ghazali that the earth rotates on its axis.

Like so many scholars in these later troubled times, Khayyam spent much of his life moving from patron to patron, unable to avoid the turmoil of the age, as rulers rose and fell and political and religious factions wrangled. No wonder that in his famous *Rubaiyat* he is fatalistic:

What we shall be is written, and we are so.
Heedless of Good or Evil, pen, write on!
By the first day all futures were decided
      (From the translation of Khayyam's *Rubaiyat* by
      Omar Ali Shah and Robert Graves)

Euclid's Fifth

Khayyam was one of the many Arab scholars, including al-Tusi and ibn al-Haitham, who tried to work out a proof for what is called Euclid's Fifth Postulate, or the Parallel Postulate. The Fifth Postulate is about parallel lines. If part of a line crosses two other lines so that the inner angles on the same side add up to exactly two right angles, then the two lines it crosses must be parallel. This postulate is at the heart of basic geometric construction, and has countless practical applications.

But it is surprisingly hard to prove. Omar Khayyam came close, but it was ultimately unprovable. Euclid's geometry works well for flat, two- or three-dimensional surfaces and most everyday situations. But just as the earth's surface is not actually flat, however much it appears to be, so space is actually curved and has many more than three dimensions, including those of time. Euclid's parallel postulate means that only one line can be drawn parallel to another through a given point. But if space is curved and multi-dimensional, many other parallel lines can be drawn. This is why mathematicians such as Gauss began to realise such limitations of Euclidean geometry in the 19th century and developed a new geometry for curved and multi-dimensional space.

Triangulating faith

Trigonometry was first developed in ancient Greece, but it was in early Islam that it became an entire branch

147

of mathematics, as it was aligned to astronomy in the service of faith. Astronomical trigonometry was used to help determine the *qibla*, the direction of the Ka'bah in Mecca. Modern historians such as David King have discovered that the Ka'bah itself is astronomically inclined. On one side it points towards Canopus, the brightest star in the southern sky. The axis that is perpendicular to its longest side points towards midsummer sunrise.

Mecca's significance is such that when a deceased person is to be buried, contemporary Islamic tradition determines that his or her body must face Mecca. When the famous call to prayer is announced, it must be done facing Mecca. And when animals are slaughtered, slaughtermen must also turn in the direction of the holy city. Islamic-era astronomers began to compute the direction of Mecca from different cities from around the 9th century. One of the earliest known examples of the use of trigonometry (sines, cosines and tangents) for locating Mecca can be found in the work of the mathematician al-Battani, which, according to David King, was in use until the 19th century.

Geometric designs

Yet another example of the interplay of mathematics and faith can be found in the decorative geometric patterns adorning some of the world's most famous mosques. These designs are known in the Western world as 'Islamic' geometric patterns and are characterised by a single, often complex geometric design which seems to endlessly repeat itself while fitting inside a confined space. Their development (and popularity) is sometimes ascribed to the fact that in early Islamic societies, drawing or painting the human form was frowned upon – especially in the context of religious buildings. The designs were produced by craftsmen often using no more than a ruler and a fixed compass, and with little formal mathematical training. However, some Islamic-era mathematicians attempted to describe the patterns using mathematics. Among them was the rationalist philosopher al-Farabi (from the 9th century), whose book on the subject is called *Spiritual crafts and natural secrets in the details of geometrical figures*. Another book by the 10th-century mathematician Abul Wafa is entitled *On those parts of geometry needed by craftsmen*.

The story of Islamic-era geometric design also helps to provide important clues to two questions. First: to what extent did the needs of religion drive the quest for science and knowledge? And second: to what extent did Muslims adopt scientific methods to help them carry out their faith obligations?

Faith and learning

Al-Khwarizmi's book on algebra as a way of calculating inheritance as prescribed in the Qur'an and al-Battani's trigonometrical solutions to finding the direction of Mecca both point to the fact that, in a limited number of cases, the needs of faith did indeed inspire feats of learning. However, when it comes to the second question, it is clear that the mass of Muslims did not feel comfortable having to get their heads around a complicated new science – indeed, on the contrary, they discovered that there were far simpler ways of pleasing God.

Yahya Michot of the Hartford Seminary in Connecticut says that an important reason why Islam became so popular in such a short space of time was that its rituals were relatively simple to carry out. They did not demand a big commitment to learn new techniques, nor did they require a mastery of complicated instruments; nor indeed did Muslims need access to higher expert authorities in being told what to do.

So, whereas al-Battani devised a clever way of determining the direction of Mecca, believers (then as now) did not immediately rush out to learn the principles of trigonometry. A much easier way of finding Mecca was to follow the practice of Muhammad, who would pray due south when not in Mecca and who is reported to have said: 'What is between east and west is a *qibla*.' Many of the earliest mosques point due south. A few were aligned along the direction of a road that might be heading towards Mecca. Others pointed towards specific walls of the Ka'bah.

Similarly, whereas Islamic-era astronomers took pains to compute accurate tables on the phases of the moon, Sunni-Islamic tradition to this day demands that a new moon need only be spotted with the naked eye for a new calendar month to begin. Theoretically speaking, complex tables are not needed. And whereas mathematicians have for centuries been working on ever more accurate tables for the times of prayer, throughout the Islamic world (especially in hot, outdoor environments), many among the faithful continue to rely on the length of a shadow to work out the times for prayers during the day.

11

At Home in the Elements

My wealth let sons and brethren part. Some things they cannot share: my work well done, my noble heart – these are my own to wear.

Jabir ibn-Hayyan, 8th century

None of the sciences practised in Islamic times had such an ambiguous reception in the modern age as chemistry. The very name of the science has many meanings. Chemistry was a field of study in ancient Egypt and also in classical Greece. Somewhere, during this time, emerged the word *kimia*, which is believed to have been modified by Arabic-speaking scientists to *al-Kimya*. However, *al-Kimya* is also the source of the word 'alchemy', the technique, so rich in mystery, which aimed to manufacture gold and silver from other metals.

In early Islam, chemistry and alchemy – just as was the case with astronomy and astrology – were not nearly so clearly divided as they are now, even though there were scientists such as ibn-Sina who were as stridently sceptical of alchemy as anyone today. Others, however, were happy to work within both traditions and none more so than the giant of Islamic-era chemistry, Jabir ibn-Hayyan, known to the Latin world as Geber.

The earliest biographical mention of Jabir comes in the *Fihrist*, the famous 10th-century biographical scientific dictionary written by the Baghdad scholar ibn al-Nadim. Ibn al-Nadim describes Jabir as a spiritual healer belonging to the Shia tradition. But other scholars were less sure, and believed that his name might have been 'invented' as a cover for others.

There was controversy in Europe, too, where Jabir's work appeared in Latin in the 12th and 13th centuries in the form of five treatises. Some historians claim that these did not come from Arabic originals, but were penned by a contemporary European who they dub 'pseudo-Geber'. Historians have delved into the linguistic turns of phrase in the treatises in order to see if this is true. Some say that the authentic Arabic phrases that can be found in the text prove that it came from an Arabic original. Others say that there were plenty of proven forgeries which deliberately used Arabic phrases to create an aura of authenticity.

When it comes to other texts attributed to Jabir, the matter doesn't get any easier. Much of his work was written in codes and symbols. It's not clear why he wrote this way. It may be that, like so many alchemists, he wrote in code in order to keep his work secret from anyone but the initiated. Alternatively, his reason for writing in code might have been to avoid the risk of being charged with heresy for his more challenging work. However, the interchangeable symbolism of numbers, letters and words is endemic to the Arabic language – and helped to give us the very profound, scientific and practical mathematics of algebra.

The real Jabir

However, two factors indisputably show that however it came about, and whoever actually produced them, these treatises represent a towering scientific achievement which helped to lay the foundations of modern chemistry. One is the Jabir texts whose origin we do know for sure, which are rich in descriptions of the basic laboratory techniques and experimental methods that are essential to chemistry. Second is the real chemistry – the key substances that were identified such as sulphuric and nitric acid, the processes that were discovered including distillation, sublimation and reduction, and the scientific equipment that appeared, such as the alembic and the retort. All these things came from somewhere, and if it wasn't a man called Jabir, then it was someone who deserves an equally high place in the history of science.

It is thought that Jabir was born in Tus in Khorasan (in modern Iran) in around 722, and that his father was a pharmacist. It may be that he got most of his chemical training from his father, but he was also living in Persia where there was a long alchemical tradition. It is said, though, that he was trained in the esoteric arts as an apprentice to one of Islam's most revered figures: Ja'far al-Sadiq. References throughout Jabir's works to 'My Master' are thought to be to al-Sadiq. Thereafter, nothing is known of him until he appears in Kufa in Iraq in the time of the Abbasid caliph Harun al-Rashid. It seems he was drawn nearer Baghdad by the Barmakids, the powerful Persian family who acted as advisors to the first Abbasid caliphs. But that Barmakid connection, though it gave Jabir the finance and the status to work on his science at the highest level, was to prove his downfall. When Ja'far the Barmakid was executed by Caliph Harun, those most closely attached to the Barmakids lost their status or, like Jabir, were kept under house arrest.

The quest for artificial life

In his work, Jabir also drew on traditions from Egypt, where there was not only much knowledge of chemical processes, but a history of esoteric masters such as Hermes Trismegistus (the 'triple master') who explored the occult nature of the relationship between different substances. This Egyptian tradition probably reached the Islamic world via ancient Greece. The other tradition he was influenced by was that of the ancient Persian Zoroastrian magi, which Jabir had no doubt encountered directly in Tus.

Jabir delved deep into alchemy. Some say his ultimate aim was not the usual quest to turn base metal into gold, but something more. It was the quest for *takwin*, the artificial creation of life, and in his writings he alludes to recipes for creating snakes and even humans. This quest was subsequently to inspire the literature of Faust and, subsequently, Mary Shelley's *Frankenstein*. No one knows, of course, whether Jabir was seriously experimenting with this, or whether his writings on this matter were symbolic.

Yet though his alchemical work, Jabir also explored chemistry in a clear, matter-of-fact, and lucidly experimental way that was both entirely new and completely his own. This is why Jabir is often described as the 'father of chemistry'. 'The first essential in chemistry', he insisted, 'is that you should perform practical work and conduct experiments, for he who performs not practical work nor makes experiments will never attain the least degree of mastery.'

The experimental method

The methods of working described in Jabir's writings are detailed, and helped to put chemistry on a scientific footing. His descriptions of how to produce certain chemicals, or perform certain processes, are called recipes, and they read like the instructions for making a cake. But they are clear enough for anyone to follow, and they establish a template for detailed meticulousness. Such was Jabir's care for precision that he invented a scale that could weigh accurately within less than a sixth of a gram. It may be that this precision led him to speculate that when chemicals combine, their basic nature is retained at a level that is far too small to see.

For Jabir, as for so many scientists, experimenting with matter meant going into his work room and seeing what happened when he mixed substances, heated them, cooled them, crushed them, baked them, stirred them and so on – the classic vision of the alchemist's den and later the chemistry laboratory. To do this with the precision he needed, Jabir used or invented a variety of flasks, such as the retort. He is also thought to have discovered various chemical processes such as reduction and sublimation and, most important of all, distillation – or at least if he didn't discover distillation, he found a way to achieve it with his invention of the alembic, a simple enclosed flask for heating liquid, with a spout for draining off the drops that form as the vapour condenses in the top of the flask.

With the alembic, wine could be made into alcohol. This was not used for making strong alcoholic drinks, of course, since Islam forbids them, but it became the key process in a number of chemically-based industries which took off in the Islamic world, including perfumery, ink- and dye-making, and producing drugs and particular chemicals. The alembic was also later used for distilling mineral oil to make kerosene, which was used for fuelling oil lamps.

Jabir is also credited with the discovery of strong acids – sulphuric, hydrochloric and nitric – which were so strong that they could could dissolve metals. Fortunately he also discovered the substances that could neutralise them, alkalis – another Arabic word that has come to us through chemistry. He also discovered the one acid that could dissolve gold and platinum: *aqua regia* or royal water, the short-lived combination of hydrochloric and nitric acids. This discovery inspired countless generations to pursue the search for the magic formula that would turn base metal into gold. But the discovery of strong acids and alkalis is more fundamental than it might sound at first. These substances are essential in modern chemistry and in the industrial chemical processes which produce so many of the things we rely on today, from plastics to artificial fertilisers.

Jabir also tried to provide a framework for classifying chemicals. Part of this came from the old Greek notion of the four elements – fire, earth, air and water – but he developed this by grouping substances into metals, nonmetals and substances that could be distilled. This is not so very different from the groups found within the modern periodic table, which identify metals and non-metals, as well as volatile substances.

Al-Razi and beyond

A century or so after Jabir, al-Razi, who was soon to be more famous for his medical achievements, began to pick up where Jabir had left off. Al-Razi refined Jabir's classifications, and distinguished between naturally occurring substances and those created artificially in the laboratory. He also emphasised the need for proof by experimentation, and refined the raw processes of distillation, evaporation and filtration.

Mineral and herbal drugs had been in use for thousands, if not hundreds of thousands, of years before al-Razi's time, but he contributed to the development of pharmacology – in which chemicals are carefully mixed in small but precise quantities and prepared to make drugs. Other scientists such as al-Biruni, al-Zahrawi and Abu al-Mansur Muwaffaq took this further, and their treatises on drugs and methods of preparing them had a major impact in Western Europe when they reached that region in the late middle ages.

Alchemy in the Egyptian-Persian-Arabic tradition continued to attract serious-minded adherents into the late 18th century – Robert Boyle and Isaac Newton were closet alchemists. But eventually a distrust of the occult – and of charlatan alchemists who promised a way to create gold – undermined its attraction so fundamentally that it waned, though has still not completely vanished. Chemistry, on the other hand, had been placed on a firm scientific foundation and today is one of the key scientific disciplines.

12

Ingenious Devices

I was fervently attached to the pursuit of this subtle science [of
machines] *and persisted in the endeavour to arrive at the truth.*
The eyes of opinion looked to me to distinguish myself in this beloved
science. Types of machines of great importance came to my notice,
offering possibilities for types of marvellous control.

Badi al-Zaman al-Jazari, Turkey, 1206

Few figures in Islamic scientific history are more colourful or intriguing than three brothers: Jafar-Muhammad,
Ahmad and al-Hasan. They lived in Baghdad in the
time of the Abbasid Caliph al-Mamun in the early 9th
century, and have come to be known collectively as the
'Banu Musa brothers'. Their father, Musa ibn-Shakir,
is reported to have been a highwayman when he was
young, but somehow he managed to put his past behind
him, becoming not only an astronomer and astrologer
but a close friend of the Caliph Harun al-Rashid himself.
He died young, leaving three small sons. Harun's son,

the Caliph al-Mamun, patron of science and rationality, made a point of looking after them.

As the young boys grew up, they were given the run of al-Mamun's House of Wisdom, and they clearly made the most of it. They were all brilliant scholars, and did much to stimulate the translation project, sending out envoys and paying small fortunes to retrieve manuscripts from the Byzantine empire and elsewhere. They quickly mastered Greek and were soon writing their own important treatments of the maths of cones and ellipses, building on the work of Apollonius. They were also accomplished astronomers, and at al-Mamun's request were able to make an accurate measurement of the earth's circumference. Yet, apart from their reputation for stirring up trouble, what really made their name was the wonderful machines and devices they created to delight the Baghdad court.

Marvellous toys

The Banu Musa may well have designed industrial or scientific machines, but if so they are lost. What we do know of their work is that they designed toys. They describe 100 of their devices in a work called the *Book of Artifices* written in 830, and each one that historians have so far examined is a masterpiece of ingenuity. Fountains that change shape by the minute, clocks with all kinds of little gimmicks, trick jugs, flutes that play by themselves, water jugs that serve drinks automatically, and even a full-size mechanical tea girl that actually serves tea. Such devices still astonish today when they are reconstructed, but they must have made al-Mamun's court gasp with wonder and delight.

Although they are just toys, the inventiveness that the Banu Musa put into them is impressive, as is the ground-breaking technology in one area of engineering: the field of automation. By making clever use of one- or two-way self-closing and -opening valves, devices for delaying action and responding to feedback, and simple mechanical memories, they created automatic systems which are no different in principle from modern machines. They used mainly water under pressure rather than electronics, but many of the operating principles are the same.

Using water to tell the time

The idea of using water pressure to achieve automation reached its pinnacle in the development of clocks. The need to know what time to pray was a crucial spur in Islam to the development of water clocks which could keep the time through day and night. Water clocks such as that of al-Zarqali in Toledo (11th century) became the wonders of the age.

One extraordinary device is a water clock in the shape of an elephant, designed by an engineer called Badi al-Zaman al-Jazari and illustrated and described in his *Book of Ingenious Devices* (1206). The elephant clock combined water principles from Archimedes with an Indian elephant and water timer, Chinese dragons, an Egyptian phoenix, a Persian carpet and Arabian figures.

Al-Jazari was born in the region of al-Jazira between the Tigris and Euphrates in the 12th century. This was a time when the Turkic-speaking peoples were already beginning to make this part of the world their own, and in 1174 he went to work for the Banu Artuq, the rulers of Amid (now known as Diyar Bakir in southern Turkey). There may have been many engineers as talented and as innovative as al-Jazari, but he was also a skilled communicator who could write and draw too. He must have been an old man, though, when the Prince of Amid, Nasir al-Din Mahmud, ordered him to write his book, for within a few months of completing it he was dead.

Researchers are just beginning to go through this book, which seems to be the culmination of Islamic mechanical technology, to try out some of these machines – either on computers or by building models according to al-Jazari's designs. What they are finding is beginning to cause quite a stir.

Technology transfer

Historians often find it hard to know precisely how important ancient technology was to the making of the modern world. In many areas of science, for example, the discovery of scientific manuscripts can help scholars to follow a paper trail which can show how ideas spread. Manuscripts also contain acknowledgements, which tell the reader who else needs to be credited with particular discoveries. This is how we know, for example, that the astronomer Nicolaus Copernicus used sources written in Arabic.

However, with technology, it's not always so easy to see how a new invention or discovery came to be. Without physical evidence we cannot be sure whether an invention was entirely the work of the inventor, or the extent to which he may – or may not – have borrowed from his peers. This is partly because many of the 'missing links' in terms of working objects from the past are simply not available.

These are some of the reasons why it's hard to be sure of the precise extent to which Islamic technology was picked up in Western Europe, and the extent to which modern technological developments came about independently of things that happened in the past. From the examples of al-Jazari and the Banu Musa brothers, it seems that technology in Islamic times was notably advanced. We find references to crankshafts, which would become a key component in the machines of the European Industrial Revolution. We find cam-operated valves, which came into their own in the internal combustion engine. And there are references to automatic valves and double-action pumps, as well as technology directed towards raising water – and also using water to provide power.

Many technologies helped to power the Industrial Revolution, and from what we have seen in this chapter, the scientists and engineers of the Islamic world could well have played a part.

Part III

Second Thoughts

13

An Endless Frontier

Whosoever seeks the truth will not proceed by studying the writings of his predecessors and by simply accepting his own good opinion of them. Whosoever studies works of science must, if he wants to find the truth, transform himself into a critic of everything he reads. He must examine tests and explanations with the greatest precision and question them from all angles and aspects.

Hassan ibn al-Haitham, Cairo, 10th century

Much of this book has been concerned with how scientists from Islamic times contributed to the modern world. We have looked at both scientific and industrial processes. The previous chapter looked at engineering, and before that we had mathematics (algebra and trigonometry) and medicine. And in astronomy, for example, Arabic-speaking astronomers were found to have made a contribution to the work of Copernicus. Islamic scientists, however, had an impact in other fields which may well have helped shape the world as we know it today, including optics and the development of universities.

Moreover, there is good evidence of early thinking in the Islamic world on the topic of human origins.

Seeing is believing

The nature of vision, and finding out the mechanisms for sight, are among the oldest questions in the history of human knowledge. These were of interest to scientists from the Islamic world too, and by the time of the Caliph al-Mamun and the translation movement from Greek to Arabic, scholars such as ibn-Sina and ibn al-Haitham were well aware of the leading theories of the time.

Perhaps the most popular of these theories of vision was what is now called the *extramission* theory, whose proponents included Plato. According to extramission theory, the human eye is able to see objects because the eye releases a special kind of optical energy. This energy can be regarded as being a bit like electromagnetic radiation; it streams ahead out of the eye in pulses, shining a sort of light, which allows humans to see.

The extramission theory wasn't without its critics, however, and they included Aristotle. The critics believed that, instead of a light pulsing out of the eye, our vision is more likely to come from a light that is released from physical objects themselves, which then interacts with the eye. This theory is known as *intromission*, and is not far off from our latest knowledge of vision.

Galen, the pioneer of herbal medicine, had yet another view: he shared the extramission idea that the eye emits optical energy, but he also held that our ability to see happens when this energy mixes with the surrounding air and with sunlight.

Among the first scientists of the Islamic world to get to grips with theories of vision was Abu Yusuf al-Kindi, the Baghdad governor's son who became scientific advisor to three caliphs, starting with al-Mamun. Al-Kindi, like his patron, acknowledged the Greeks as the original masters but, like many of his contemporaries, he also knew that advances in learning would require improving and refining ideas from the past:

It is fitting for us to remain faithful to the principle which we have followed in all our works, which is first to record in complete quotations all that the ancients have said on the subject. Second, to complete what the ancients have not fully expressed, and this according to the usage of our Arabic language, the customs of our age and our own ability.

(From *Theories of Vision: From Al Kindi to Kepler* by David Lindberg, Chicago, 1976)

Other proponents of extramission from the Islamic world included al-Farabi (who died in 950) and the astronomer Nasir al-Din al-Tusi.

Extramission was also supported by a second group of scientists, led by the medical doctor and translator from Baghdad, Hunayn ibn-Ishaq. Hunayn and members of his camp were largely on Galen's side of the argument, however. They believed that, yes, the eye emits an optical energy, but that our ability to see is achieved when the energy from extramission mixes with air and sunlight.

Criticising the rival intromission theory, Hunayn asked his readers to imagine that a large group of people – say 10,000 – are all standing before a tall mountain. If this mountain were capable of emitting images of itself, Hunayn said, then it would need to know that it has to send out 10,000 sets of images so that every person standing before it can see the mountain. As it is impossible, he argued, for mountains to know how many people will be looking at them, this meant that intromission had to be false.

Hunayn's ideas were extremely influential, both in the Islamic world and beyond, according to historian David Lindberg. Hunayn's book *Ten Treatises on the Eye* 'influenced almost every member of the Western optical and ophthalmological tradition before the 17th century'.

A different view

Intromission, though less popular among Islamic scientists, did have influential supporters. Among these was al-Razi (who died in 924). Using his experience as a working physician, al-Razi discovered that the pupil of an eye contracts and dilates depending on how much external light it receives. This was in stark contrast to Hunayn's view that the pupil dilates according to the pressure of the visual energy that the eye is about to release.

In addition to al-Razi, some of the most powerful and convincing attacks on extramission were made by ibn-Sina. In spite of being Galenist in his medical writings, ibn-Sina parted company with his mentor on extramission. Extramission was contrary to common sense for ibn-Sina; he could not believe that something as small as an eye could produce the energy necessary to be capable of travelling great distances, such as to the stars in the sky, and that this process would need to happen every time the eye opened. Moreover, said ibn-Sina, if it were true that vision occurs when energy from the eye mixes with air and sunlight, then stars and distant planets should be invisible to the naked eye, because air does not touch these distant objects.

Ibn-Sina's critiques of extramission were powerful and to a certain extent convincing. However, he was unable to significantly advance our understanding of vision. Instead, the job of taking the study of optics to new heights fell to ibn al-Haitham.

Optics goes to new heights

As we have already encountered, ibn al-Haitham lived in the 10th century and worked for the Ismaili caliphate (the Fatimids) based in Cairo, under the ruler al-Hakim. He worked in a range of fields, though he is known in the West chiefly for his works on optics and astronomy, including *The Book of Optics*, *On the Spherical Burning Mirror*, *On the Light of the Moon*, and *Doubts Concerning Ptolemy*. Ibn al-Haitham was a skilled experimentalist and he used his abilities to great effect when testing out the theories of the day.

He began his criticism of extramission by describing what happens when people are exposed to bright lights. For example, anyone who tries to look directly at the sun experiences pain, he said, as do those who try to look at the sun's reflection in a mirror. No matter what the light source, the effect of bright lights, according to ibn al-Haitham, was always – and painfully – the same. This suggested to him that light entering into the eye from an external source had at least some role to play in vision.

Furthermore, he argued, even if we did accept that the eye released a visual energy which interacts with the air (Galen's view), the result of this interaction would need to flow back into the eye so that vision could be regis-tered by the observer's brain. This confirmed to him that, even if we accept extramission, some form of intromis-sion would be needed for the eye to be able to see.

To test his ideas out further, he began to experi-ment with refraction, which is the bending of light as it passes from one medium to another. According to ibn al-Haitham, if vision is what happens when light passes from an object and into the eye, it is likely to bend once it enters the eye. This refracted light could lead to a dis-torted image, so ibn al-Haitham performed many experi-ments to see if it was possible for light to transfer from one medium to another without being bent.

Ibn al-Haitham's other main contribution to optics was in suggesting that the mathematics of optics – such as reflection and refraction – need to be consistent with what we know about the biology of the eye. This was all groundbreaking stuff, and his theory of vision was enormously influential. And, as was often the case, it was more influential among Western scientists than those in his own region. Our current understanding of vision did not come directly from ibn al-Haitham, but what is not in doubt is that he was among the first to demonstrate critical flaws in the extramission theory.

Back to school

The consensus among historians is that the astronomer from Germany, Johannes Kepler, to whom we owe much of our knowledge of optics and astronomy, leant on ibn al-Haitham's work, which was widely available in Latin in the 16th and 17th centuries.

By the 16th century, scientists in Western Europe such as Kepler would have likely studied in or worked at universities. Universities in Western Europe began to appear in the middle ages. The university of Siena, for example, founded in 1240, is one of Western Europe's oldest seats of learning. Along with Bologna, Cambridge, Oxford, Padua, and Europe's oldest university in Paris, it helped to revive knowledge and learning in Europe after the middle ages. Yet, to any modern-day visitor from the Middle East or from Asia, the architecture of the oldest universities in Europe mirrors that of many Islamic-era colleges, the first of which were established in Baghdad in the 9th and 10th centuries, and later in Cairo, Egypt. Perhaps the most obvious architectural feature is the presence of a large square or rectangular courtyard, surrounded by teaching rooms at the perimeter.

The similarities, however, run deeper. Some historians, notably the late George Makdisi, have found intriguing connections between the organisation of learning in Western Europe and that in similar institutions from the Islamic world. Makdisi, for example, discovered that some of the words and concepts commonly used in modern higher education and scientific research today have a connection to an Islamic past.

Principal among these is the 'doctorate', a concept whose origins date back to the time of Europe's earliest universities. The origins of the doctorate, however, are believed by historians such as Makdisi to lie even earlier still, in a certificate or diploma called a 'licence to teach and issue legal opinions'. This was awarded by senior teachers in the Islamic world's colleges to those of their trainees who could demonstrate, after a number of years of study, that they had absorbed enough knowledge to be able to teach their own students. Makdisi discovered that teaching diplomas were used for the same purpose in Bologna and Paris two centuries later.

Similarities notwithstanding, Europe's first universities are different from Islamic-era colleges in one important respect. Universities such as Bologna, Oxford and Paris were established with the support of politically powerful churches, and their aims included the need to train new generations of scholar-clerics who were expected one day to hold the reins of power.

Islam's first colleges, in contrast, grew out of a movement against state-organised religion – and they were
not places where the leading scientists of the day came
to work or to study. Almost all of the scientists we have
encountered in this book worked directly for caliphs and
governors, and were often based inside palace complexes.
We have seen that one of these caliphs, al-Mamun, initiated an inquisition, ordering the persecution of intellectuals who refused to accept rationalism within Islam.
What is less well known is that once the inquisition failed,
those who had resisted the caliph's demands decided to
organise themselves (in the form of guilds) so that they
could resist future attempts at governmental meddling
in religious learning and interference in who should – or
should not – have the right to become a teacher and have
students.

These guilds later gave rise to the first colleges, and the
'licence to teach' was designed both to increase the numbers of scholars who could stand up to the state and at the
same time to create a curriculum that excluded subjects
such as philosophy (and possibly the natural sciences),
which would have been identified with al-Mamun's policies to enforce a state religious doctrine.

That is not to say that different Islamic caliphates
did not continue to interfere with teaching and learning
to pursue their own goals – including science and philosophy. This would still happen, and examples include
a network of institutions set up in Baghdad in the 11th
century by Nizam al-Mulk. These, incidentally, were
partly established to counter what the leader saw as the

threat from the Ismailis and the Fatimid caliphate. Al-Azhar university in Cairo had already been established by the Fatimids, in part so that they could train their own scholars and theologians in their own more rationalist doctrines.

If the origins of the doctorate are one day found to lie in the ancient cities of the Middle East, one inescapable conclusion from this is that a key component of our modern research enterprise has roots in two seemingly contradictory aims: the first a desire to free scholarship from state control; the second a desire to stop young people from innovating and experimenting with ideas, and instead to steer them in the direction of traditional thinking.

A second conclusion might be even more troubling. As we shall see in the next chapter, Europe's nations used learning – both scientific research and higher education – in the service of colonisation. If Islam's colleges were the forerunners to Europe's universities, could it be said that Islamic-era science and learning had a small part to play in the colonial project?

Acknowledging the past

As can be seen in this chapter and elsewhere in the book, scientists from the Islamic era were generous – perhaps a little over-generous – in acknowledging that their knowledge of optics, astronomy, medicine and much else had in fact been developed elsewhere, especially India and ancient Greece. What the Western world calls Arabic numerals are known as 'Indian numerals' in the Arabic language, and what is known in the Western world as Islamic medicine is known in Muslim countries as Greek (or *Unani*) medicine.

Yet, when it was Europe's turn, not all in the research fraternity were willing to repay the compliment and acknowledge that some of the ideas being worked on in the 15th and 16th centuries had come to Europe from the empires of non-Western cultures. Such a lack of citation was not universal, and it proved less of an issue in the fields of optics, algebra and chemistry. As we have seen, controversy is greater over the extent to which Islamic-era astronomy, medicine and the organisation of learning was adopted in Europe.

In his book *Islam and the Destiny of Man*, the British writer and former diplomat Charles Le Gai Eaton argues that the present time (in which the nations of Islam and the post-Christian world are mostly at peace) is unusual. For most of the past, he says, the relationship between the two has been characterised by wars and by mistrust. These wars span literally centuries of conflict between Islam's early caliphates and Byzantium, followed by centuries of Crusades, and followed after that by centuries of fighting between European states and the Ottoman empire, right up to the 20th century and the end of the First World War.

Such a long and sustained history of warfare may offer one explanation as to why the institutions of Europe would have been reluctant to acknowledge the validity of (or cite) Islamic-era learning. One way of understanding this relationship is to consider what happened during the Cold War and its aftermath. Even if they had wanted to, scientists from East and West weren't able to acknowledge each others' work because of the deep hostilities that existed between the United States and the former Soviet Union.

A more relevant example, however, can be found in the reception of the medicine of ibn-Sina in Renaissance Europe. The *Canon of Medicine* has arguably had the greatest worldwide impact of any medical textbook in the pre-modern world. It dominated medical teaching and research for more than five centuries, and in doing so it changed how medicine was performed in Latin Europe. At the same time, however, ibn-Sina was sometimes harshly attacked by commentators in Western Europe, both for his medicine and also for the fact that he was from a different religion.

Much science and medicine arrived in Europe in Arabic through Spain – specifically via a translation school in the city of Toledo. But quite how the *Canon* got into university curricula is not clear. The first known Latin translation was by Gerard of Cremona in the 12th century. This translation became the standard text in Western Europe – very few saw any reason to go back to the Arabic original after that.

The *Canon* was soon being taught to budding medics in France, Germany, Italy and Spain at universities including Bologna, Montpellier, Padua, Paris and Tübingen. Thanks to the painstaking work of New York-based historian Nancy Siraisi, we know of the existence of 60 separate editions that were in circulation between 1500 and 1674.

However, by the mid-14th century, the critics had begun to make their presence felt. By the 15th and 16th centuries, the attacks on 'Avicenna', and on those who defended his work, were indiscriminate and sweeping. The attacks were of three kinds. There were the traditionalists, those who felt that true medicine resided in Greek thought and that ibn-Sina had got Galen wrong. Second were those who felt that the Christian West had no reason to learn medicine from a non-believer. This was the time of the Crusades, of course. A third group of critics were those scientists who thought the book to be out of date, and who argued that its method of teaching was of little use in hospitals and surgeries. Refreshingly, the critics didn't ban the book: they published their own commentaries alongside the Latin text.

In the 14th century, the medical writer Francisco Petrarch described the *Canon* as 'Arab lies'. Bassiano Landi, a 16th-century professor of medicine at Padua, lamented how his predecessors had been 'misled by the bad leadership of the Arabs'. In 16th-century Germany, medical professor Leonhart Fuchs said: 'the Arabs had taken all their knowledge from the Greeks and, like Harpies, defiled all that they touched'. In 16th-century France, Symphorien Champier accused ibn-Sina of being part of 'that filthy and wicked Muhammadan sect, which legitimises divorce and takes the view that all miracles have a natural explanation'.

Fortunately, ibn-Sina had robust defenders too. Girolamo Cardano, a distinguished professor at Bologna in the 1500s, said that ibn-Sina was arguably the greatest medical practitioner since Hippocrates. It was laughable, he believed, to criticise ibn-Sina for being a Muslim when Galen worshipped idols; and no religion, according to Cardano, had a monopoly on science or philosophy. Furthermore, Benedetto Rinio, a physician from Venice, said that it was absurd to attack ibn-Sina for drawing on the work of his predecessors – when this is exactly what Aristotle and Galen had done.

Where did we come from?

One less-known field of knowledge which Islam's scientists explored is the area of human origins. Two questions in particular exercised much thinking and searching for answers: where did we come from, and what happens when we die?

As you would expect, their starting point was the Qur'an, which contains a welter of material on human origins in particular. But like any religious text, it can be interpreted in many ways, and this allowed scientists and philosophers to suggest alternative explanations to the story that is still told to millions of children and adults.

Like the biblical texts, Islam teaches that God punished Adam and Eve for falling into Satan's trap and eating from the forbidden fruit tree. Muslims believe that the world will eventually come to an end, after which there will be a judgement day in which humans will rise again and be called to account for their time on earth. The righteous will live forever in a place called heaven, while sinners will burn forever in hell. Islamic texts, however, differ in that Adam and Eve were later forgiven by God and were told to create life on earth, which was always part of God's plan.

Most interestingly in terms of science, the Qur'an says that God created humans in 'stages'. Several verses also talk about the aquatic origins of life. Islamic teachings such as these provided scientists with room to speculate about the nature of human origins and what happens at the end of life.

Speculating about evolution

Historical speculation about human origins is not new, and the Islamic world, like many of the world's cultures, presents a long paper-trail of thinking and writing on the subject.

The first documented example is from someone who could well be the Islamic world's first professional science-writer – as he earned a living from writing about science. This was al-Jahiz, who was from East Africa but moved to Baghdad in the 9th century and was known to the Caliph al-Mamun. The most famous of his some 200 books is *The Book of Animals*, in which he describes the characteristics of 350 different varieties: 'Animals engage in a struggle for existence [and] for resources, to avoid being eaten and to breed.' He continues: 'Environmental factors influence organisms to develop new characteristics to ensure survival, thus transforming into new species. Animals that survive to breed can pass on their successful characteristics to [their] offspring.'

Another example can be found in a 10th-century text called *The Book of the Yield* by Muhammad al-Nakhshabi, an Ismaili thinker from Central Asia, who wrote: 'While man has sprung from sentient creatures [animals], these have sprung from vegetal beings [plants] and these, in turn, from combined substances, these from elementary qualities, and these [in turn] from celestial bodies.'

Later writers to speculate about evolution included the 13th-century poet Jalaluddin Rumi and the early-20th-century philosopher Muhammad Iqbal, who was also Pakistan's national poet. Iqbal had read Darwin and got into a debate (using the Qur'an in support of his position) on the issue of whether or not humans are *still* evolving. For Iqbal, God could not possibly have created human life and then left it stagnant. Improvement, modification and innovation in human life and consciousness were all part of God's plan for variety and diversity, in his view. 'There is nothing more alien to the Qur'anic world,' he wrote, 'than the idea that the Universe is a temporal working-out of a pre-conceived plan; an already completed product, which left the hand of its maker ages ago and is now lying stretched in space as a dead mass of matter to which time does nothing.'

Iqbal believed strongly that human evolution has not come to an end, though his reasons for thinking this were rooted firmly in the realms of faith, and were based on the idea that there is such a thing as a perfect human.

Muslims are taught to emulate the Prophet Muhammad, who is seen in many Islamic traditions as the perfect man. He argued that it was the will of God that humans could one day achieve such perfection in carrying out God's work on earth, so that they could become close to the ideal of Muhammad. Elsewhere, in a verse of poetry, Iqbal challenges God to improve on what he sees as the inferior qualities of the human race, especially the tendency towards cruelty and meanness:

> Design a new pattern
> Create a more perfect Adam
> This making of playthings of clay
> Is not worthy of God, the creator
>
> If the pattern is poor
> What does repetition achieve?
> How can the cheapness of man
> Meet your approval?

(From *Iqbal's Educational Philosophy*,
by K.G. Saiyidain, 1938.)

14

One Chapter Closes, Another Begins

As the story so far has clearly demonstrated, manuscripts show how scientists from the Islamic era were experimenting, innovating and pushing the boundaries all the way up to the 16th century. After that point, however, the records for this kind of work start to become thin. Not only that, but bricks-and-mortar evidence of scientific activity also proves hard to come by. Travel today to the former capitals of the Islamic era, such as Baghdad, Bukhara, Cairo, Damascus and Istanbul, and, with a small number of exceptions, you will be pushed to find evidence of the great institutions that leap out of manuscripts and history books: the observatories, hospitals, schools and colleges which so much of this book has described. A centuries-old tradition of learning seems to have largely disappeared.

Many observatories and hospitals, for example, stand as neglected ruins. In some cases, the destruction was so complete as to leave no trace. In other cases, former institutions of science and learning are now sites of national heritage. Why what was once a working hospital or an observatory should become a candidate for a heritage site is a key question for historians of Islamic science. It's one that can best be answered by looking more closely at two related issues: the final years of the last two Islamic empires, the Mughals and the Ottomans; and the traumatic experience of colonisation, whose seeds were planted around the same time as advanced Islamic-era science was coming to the end of its life. Thinning evidence of advanced science during the Islamic era coincides with the final centuries of Islamic rule and the rise of Western European nations as military and trading powers.

Searching for a new science

The principal players in Europe's military and trading encounters with Eastern nations included Austria, Britain, France, Holland, the Papacy, Portugal, Venice and Russia. Between the 15th and 20th centuries, these nations would eventually come to control, or exert a strong influence over, many of the countries that were once ruled by the Mughals and the Ottomans. The Mughals of South and Central Asia ruled from the early 1500s to 1857, just ten years before Britain formally declared India to be a British colony. Ottoman rule lasted longer, from 1281 to 1922, and came to an end partly as a result of their decision to back Germany during the First World War.

The collapse of the Islamic empires was regarded by Muslims the world over as a hammer-blow. For centuries, generation after generation had grown up being (even if only nominally) part of an institution to which they had both a physical and a spiritual allegiance. And then it had all gone. It was the equivalent of both the Church and the monarchy suddenly (and violently) coming to an end in Britain. The sense of loss was very deep and remains to this day. In 1919, the Muslims of colonial India even launched a movement to re-establish the Ottoman empire, as did (in the early 1950s) political activists in Palestine.

Perhaps surprisingly given this great sense of loss, a century before the end, senior figures in the Ottoman empire were in fact preparing to give it all up. They were tired of having to run a costly empire that many no longer believed in, and of having to fight wars on many fronts. Because of this, the Ottoman elite had by the late 1800s begun to reach out to Western Europe. They had forged friendly links with Britain and France, and were impressed by what they saw. Ottoman ambassadors in Paris and London sent dispatches home about the new museums and scientific societies, and how these were far ahead of anything inside their own borders. One group of Ottoman diplomats who were based in Europe pushed for a series of liberalising reforms to their own constitution. These included equality before the law; guarantees of certain fundamental rights, such as the right to life, property and public trials; a promise of elections to local councils; and the development of new legal codes to harmonise trade with Western Europe. All were to be written in a way so as to conform to Islamic principles. 'Europeanization will in no way reduce the value and importance of our religion', wrote one commentator, Ahmed Hilmi, who supported the reforms. 'In fact through Europeanization we may be able to resurrect the ancient civilization of Islam.'

However, not everyone wanted change. Yes, the empire was much weakened after years of conflict, and yes, there was little money in the treasury. But these were not considered strong enough reasons for what many saw as capitulation before Western powers. According to the historian Halil Inalcik, the critics agreed that the Ottoman state needed more Western science and technology to improve its prowess on the battlefield and to help improve living standards, but that was still to be at the exclusion of Western laws and Western culture. The former would be enough, they argued, to make the empire strong again, and need not violate Islamic laws. After a long and very public debate, the critics had their way. And by the late 1870s, broader reforms to social and political institutions were put aside and efforts to import Western science and technology were stepped up. What the Ottomans did next was quite breathtaking even by today's dizzying speed of technological change.

Up until the mid-1800s, Ottoman society was seen in today's terms as being pre-modern. There were few roads, no trains, little electricity, no phones. The system of medicine was still the same as that in ibn-Sina's *Canon of Medicine*. Yet, within a generation an incredible transformation had occurred. New hospitals that treated infectious diseases using vaccines based on the latest microbiology were created in 1862. The metric system of weights and measures came seven years later, and time zones were changed to Greenwich Mean Time at the turn of the 20th century. A network of post offices came in the 1830s, telegraph lines in the 1850s, and telephone lines in 1881. A railway linking Istanbul with Mecca in Saudi Arabia – the Hijaz Railway – was built between 1900 and 1908, and Istanbul saw its first school of aviation in 1912, run by pilots and engineers from France (though they were recalled by France at the outbreak of the First World War). Popular science, too, saw a boost. When the Ottoman Scientific Society opened in Istanbul in 1861, its first lecture was on modern physics and included a demonstration of experiments with electricity. A capacity crowd of 400 turned up and late-comers had to be turned away. In publishing, 28 science books were printed between 1727 and 1839. Between 1840 and 1876 that figure increased to 242.

The Ottomans had been using Western and other European technology long before the 19th century – firearms, clock-making, the magnetic compass and the printing press all found their way to Istanbul and beyond. But much of this earlier adoption of modern technology was aimed at improving the military. In 1773, for example, the Imperial School of Naval Engineering was created under the supervision of a French officer, Baron de Tott, to provide cadets with an education in modern science and engineering, but alongside more traditional subjects such as Arabic and religion. In 1806, a military school of medicine opened, teaching in French and Italian and using textbooks from Western Europe. In 1834, a whole new Imperial Military School opened on the model of France's École Militaire.

What is clear, as detailed by Ekmeleddin Ihsanoglu, a historian who studies the transfer of Western technology to the Ottoman world, is that each of the above illustrations shows that Ottoman rulers believed that they could just purchase 'black box' applied science and technology 'solutions'. As a result, their understanding of what was really needed to create rail networks and postal systems was superficial – they were interested in science and technology only as an instrument for getting things done rather than as a form of acquired knowledge; 'knowing something from first principles' was not a phrase they would have appreciated. Almost by extension, they were not interested in adopting mass higher education, nor were they inclined to create institutions of learning so that their own scientists and engineers could build on the new knowledge and carry out research and development for themselves. This view may have been reinforced by their mistaken belief that encouraging research and inquiry amounted to an adoption of Western culture, which therefore had to be opposed.

Education reforms

Attempts to encourage greater – and friendlier – contacts with Western Europe meant that by the year 1900, the Ottoman state had given international organisations permission to run 702 primary and secondary schools. Of these, 465 – the single largest share – were led by missionaries from the US, and 100 of these schools had been established during the past twenty years alone. American schools were so popular with parents that in Anatolia one in three school-age children was enrolled at such a school. Why were they so popular? One explanation seems to be that they were not just educational institutions; through these schools, children and their families were able to access the modern hospitals, pharmacies and printing facilities which the schools had established alongside their teaching function. Yet this presented a dilemma for the rulers. They wanted a degree of foreign influence in their educational system, but they did not want the system itself to be taken over by Washington, something which seemed to be in danger of happening. In one official report, for example, the minister for education described the American schools system as an 'epidemic disease'.

The government felt that it had to act. It would have liked to close the schools down, but recognised that this would lead to serious diplomatic problems. Instead, it ordered the schools to reapply for permission to teach. In addition, the American schools were told that they could no longer recruit Muslim students, nor could they locate their premises in areas where Muslims were the majority community. After much foot-dragging, the schools agreed to reapply for permission to teach, but they did not stop enrolling Muslim pupils. When the US government was pushed on this, it replied that the US, like France, Britain and Russia, had millions of nationals who were Muslims; it would change its enrolment policy only if all other foreign schools did so, too. America was too big a power for the Ottomans to mess with, and the matter was quietly dropped.

Evidence of the Ottoman state's ambivalence about Western education can also be found in its approach to building universities, known in Turkish Arabic as *darul funoon*, or home of the sciences. It took four attempts and 37 years before the first university was established in Istanbul in 1900. On the first attempt, the university opened in 1863 but closed abruptly after two years when its building was taken over by the ministry of finance. It reopened in a new building in 1869, but had to close again three years later – due this time to a combination of poor organisation and bad luck. There was a shortage of both teachers and books; many of the 450 students, who had come from traditional Islamic schools, known as *madrassas*, found the curriculum too difficult, and some were unable to afford the fees. To make matters worse, the university found itself at the centre of an ill-timed controversy over an invited lecture on the Prophet Muhammad. A famous reformist Islamic theologian from Egypt had been asked to speak during the fasting month of Ramadan. The topic he chose was what he called 'the *art* of being a prophet'. As most Muslims believe that God creates the institution of the prophet, calling it an 'art' produced a volley of complaints to the government office of religious affairs. After this incident, rumours began to circulate that the university would soon be shut down.

Classes reopened at the third attempt in 1874 and the university stayed open for seven years, managing this time to graduate one entire class. The administrators had learnt some lessons from the previous two experiences; on this occasion the university had no faculty of science, and when it opened there was no publicity and no formal launch event. The students appear to have been taught in secret, and a public announcement was made only when they entered their final year of study in 1876. The university struggled on for five more years before it closed its doors in 1881. The reason this time appears to have been a reluctance on the part of the state to continue to fund it. The university that did eventually succeed opened nineteen years later in 1900, when the state finally became serious about science and higher education.

The price of freedom

When faced with how to react to modern science and technology, India's Muslims in the British empire had much the same dilemma as those in the Ottoman territories – with the added complication that the new learning was greeted with perhaps even more scepticism because of its links to an occupying power.

Thanks to a mountain of archive material, the British empire, like that of the Ottomans, is one of the most studied in the world. The story of how, within the space of 150 years, a mercantile operation became an imperial one is well told. The first traders from Britain arrived in 1616 and sought permission to buy goods and export them back home. They also sought permission to carry arms. A key concern from the earliest days was language. The Mughal empire had used Persian (and to some extent Arabic) as the language of trade and government business. The new traders needed to learn Persian to do business with the Mughal state, and they continued to use Persian even after large parts of the country had begun to be annexed by the mid-1700s. Teaching Persian to the few hundred officials who ran a trading company was relatively easy. But by the late 1700s, many more had begun to travel to India to work as judges and magistrates, lawyers, doctors, teachers, traders and tax collectors. The Court of Directors of what was called the East India Company refused to provide the necessary tuition for such large numbers of learners, and knew that an alternative had to be found – and it came in the shape of a young surgeon from Scotland.

A lost science

In 1782, an enterprising 23-year-old from Edinburgh, John Gilchrist, had arrived in Bombay to work as a surgeon. Unusually for the time, he spent many of his spare hours walking the streets and meeting people, and even grew a beard and changed to wearing Eastern clothes so as to blend in. Gilchrist had been told that Persian was India's main language, but he quickly discovered that none of the people he was meeting could actually speak Persian or Arabic very well. At the same time, he found that his new acquaintances all had some knowledge of a language he called 'Hindoostanee'. Gilchrist had in fact discovered what today is Urdu. This is one of Pakistan's major languages and is still spoken by older generations of India's Muslims. The existence of this language was known to some in the East India Company – it was referred to as 'Moors', or 'Jargon'. The genius of Gilchrist was in how he recognised that 'Moors' could become the new language of administration for Britain in India.

Gilchrist found that if someone needed an official document in Persian, he would go to the nearest government office and explain his request (in Urdu) to a translator sitting outside, who would then produce the Persian translation. When Gilchrist tried to learn Urdu, he found that there were no textbooks, no dictionary and no written grammar. He then set about, in his own time and at his own expense, writing a dictionary. This he did by collecting a group of people he understood to be fluent in the language. Gilchrist sat down with them for hours at a time and, using an English dictionary, picked out words and asked them to describe the equivalent in Urdu. *The Hindoostanee Grammar and Dictionary* was published in 1786, and Gilchrist's efforts eventually led to Urdu replacing Persian and Arabic as the language used by Britain to administer its Indian territories.

The promotion of Urdu as India's main language for administration was something of a mixed blessing. It gave the majority population – Muslims and Hindus – immediate access to the new rulers, and undoubtedly created a future for Urdu, which it might not otherwise have had. But at the same time it gradually had the effect of cutting off future generations from the original sources that had recorded learning and science during the Mughal period. Today, very few people in India, Pakistan or Bangladesh have any knowledge of Persian. And as a direct consequence, much less is known about the history of science during the Mughal period than during other Islamic empires.

The promotion of Urdu was not the only policy decision that widened the gap between India and its Mughal-era scientific heritage. An arguably more damaging decision was made in the late 1700s when the colonial administration decided to ask educational institutions to pay rents for the first time. Throughout the Mughal period, all institutions of learning had been allowed to live rent-free. This helped education to spread, and most villages had at least one primary and one secondary school. Tax collectors from Britain, however, took a different view and insisted that the schools needed to pay. One official went so far as to describe the previous arrangement as 'a long leaky pipe' on account of the fact that a stable and regular source of revenue for the state was not being tapped. The vast majority of schools had no way of paying and subsequently closed. In the mid-1800s, an official who had been sent to do a schools survey in Madras reported that out of one million school-age children, he calculated that just 7,000 were at school. 'In many villages where formerly there were schools, there are now none.'

The Queen's English

India's Muslims now appeared to be caught in a bind. From one direction, the institutions of science and learning were being pulled from under their feet. From another, an internal debate was raging in their communities regarding the merits, or otherwise, of learning to speak English.

In addition to making profits for their shareholders, some of the empire's administrators had begun to believe that their mission was also to bring enlightenment and modernity to India. Some, such as Thomas Babington Macaulay, a member of parliament, once famously said that a single shelf of a good European library was worth more than the entire literature from India and the Arab world. Charles Trevelyan, India's minister for finance from 1860 and 1865, went further and said: 'The peculiar wonder of the Hindu system is, not that it contains so much or so little true knowledge, but that it has been so skilfully contrived to arrest the progress of the human mind. To perpetuate them is to perpetuate the degradation and misery of the people. Our duty is not to teach, but to un-teach.'

As a result, in 1835, Macaulay engineered a decision in which Arabic and Persian schooling would be phased out, to be replaced by English and modern science. A small number of the Muslim elite saw little or no conflict between being Muslim and learning English and modern science, studying at Western institutions, and retaining the ability to be critical of the colonial presence. They included the philosopher Muhammad Iqbal and the barrister Muhammad Ali Jinnah, along with others who would coalesce into the team that created Pakistan.

Engaging with the West

One prominent Islamic reformer, Sayyid Ahmad Khan, established a scientific society aimed at Muslims, and later a university, modelled on Oxford and Cambridge, that would impart both traditional and modern learning. He argued that looking to the West for knowledge of modern science was no different to when Muslims from the 9th century translated the works of Galen and Aristotle. 'And the Greeks didn't even believe in God', he told his critics. Other advocates of engagement with the new science included the Zoroastrian and Ismaili communities. When Pakistan eventually became independent, some of the institutions of modern learning it inherited, such as schools, hospitals and engineering colleges, had in fact been established by philanthropists from these communities – the descendants of much earlier dynasties and empires, for whom science and learning would always be important and would not be seen as a threat to their cultures or way of life.

But others – perhaps a majority – couldn't bring themselves to see the new Western-style education in the same light, albeit for different reasons. Some, such as Nawab Aliuddin, the Muslim ruler of one of India's feudal states, felt that the new science was a path to godlessness:

And do you suppose, sir, that I would put the evidence of one of your *doorbeens* [telescopes] in opposition to that of the Holy Prophet? No, sir, depend upon it that there is much fallacy in a telescope. It is not to be relied upon. I have conversed with many excellent European gentlemen and their great fault appears to me to lie in the implicit faith they put in these *telescopes*. They hold their evidence above that of the prophets – Moses, Abraham and Elijah. It is dreadful to think how much mischief these telescopes may do. No, sir, let us hold fast by the prophets. What they tell us is the truth, and the only truth that we can entirely rely upon in this life. I would not hold the evidence of all the telescopes in the world as anything against one word uttered by the humblest of the prophets named in the Old Testament or the Holy Qur'an.

(*A Moral Reckoning*, Mushirul Hasan, Oxford, 2005)

Others felt that anyone who advocated more knowledge of the West was effectively a mouthpiece for the new rulers, and therefore was someone not to be trusted. They regarded Sayyid Ahmad Khan's analogy with 9th-century Baghdad to be misleading, for the simple reason that Galen and Aristotle never forcibly ruled over

Baghdad in the way that Britain was ruling over India. It was not that they were against new knowledge per se; what they objected to was its direct association with an occupying power.

Plagued with problems

One of the most persuasive examples of this interplay between politics, science and religion in India at the time is in the difficulties encountered by the colonial authorities in their attempts to control and eradicate plague. In late 19th- and early 20th-century India, plague was a real killer disease, sometimes causing more than a million deaths per year. Microbiology was already an established field, scientists had isolated the bacterium that caused plague and a vaccine was available in India. Britain wanted to vaccinate the population of India, but one major obstacle it faced was that modern medicine was barely heard of, still less in use, in India. Ayurvedic medicine was the dominant force and most Muslims still used *Unani* medical therapies – prescriptions based on a combination of ibn-Sina's *Canon of Medicine* and the more faith-based *Medicine of the Prophet*. Colonial authorities knew they needed the support of traditional healers, *hakims*, to help isolate infected populations and also to encourage people to get vaccinated. But most refused to get involved, meaning that the inoculation campaigns had little impact and more lives were lost. They refused for a mixture of reasons, including the belief that a foreign power was trying to interfere in the practice of their faith.

The idea that plague was God's wrath on India was widespread. The view was that God had become angry because society had become immoral, as demonstrated by the fact that punishments for acts such as adultery were relatively light and not at all in accordance with what God had prescribed in sacred texts. Moreover, the placing of infected populations in quarantine also put new medical knowledge in direct opposition to what people believed were Muhammad's teachings. The advice attributed to him is that people should stay where they were if they came into contact with plague. One healer, Muhammad Sufi, went so far as to say that it was 'inhumane' to isolate people, especially women, from their families. On a more basic level, some other *hakims* were genuinely afraid that the content of the vaccine might itself make someone ill, or lead to loss of virility, blindness, even death.

But perhaps more fundamentally, *hakims* didn't regard plague as being contagious. Instead, they understood it to spread through 'bad air' rising from the earth and entering the body through the nose, after which it would pass on to the liver, brain and heart. Signs that the air had become infected with plague included animal behaviour, especially the death of rats. As *Unani* medicine regarded the liver, brain and heart as being the body's vital organs, early treatment was seen as being crucial to saving lives.

However, it wasn't the case that all traditional healers kept a distance. Some of the most established practitioners saw the offer to help as an opportunity to improve their knowledge of the new healthcare practices and in particular to learn alternative explanations for the spread of disease. But according to the work of Guy Attewell, a historian of medicine in South Asia, it was true that these *hakims* were in a minority.

The language of empire

Anyone who studies the British and Islamic empires will discover that, for their many differences, they also shared a few things in common. For example, both empires ruled in many of the same territories. They both – to varying degrees – encouraged converts to their respective religions. And in both cases we find examples of the use of science and technology to meet an expanding empire's many and complex needs.

In Britain's case, many of the leading lights from the Victorian scientific era cut their teeth working in India on projects that had commercial or political aims. Botanists were employed to search for rare plants with commercial potential. Surveyors and geographers were given the task of creating accurate maps, and philologists were tasked with understanding the many languages spoken by India's diverse peoples. One of the most comprehensive studies ever to have been done was George Grierson's monumental, nineteen-volume *Linguistic Survey of India*, published at the beginning of the 20th century. Finding or creating a common language is clearly central to empire-building. The Islamic empires opted to promote Arabic, and the consequences of this for other languages are still being discovered. Britain, on the other hand, chose first Urdu, then English in India and in the rest of the empire. This was a decision destined to have both good and bad consequences for science and learning in the countries that were once colonies. Ending teaching and learning in languages such as Arabic and Persian would cut off the majority of younger generations from their scholarly heritage. But at the same time, providing education in European languages would give – albeit to a minority of wealthier people – access to the latest science and technology and the tools to pursue advanced science and higher education in the countries of Western Europe, which many did.

A number of those who ran the empires of Britain and of Islam's early years have something else in common. In each case, examples can be found where force or coercion was used in the desire to create societies based on science and reason. And where this happened, the reaction from ordinary people – though not scientists – was to recoil and withdraw. The Abbasid caliphs tortured those who openly disagreed with their project to install rationalism as the religion of the Islamic state. British attempts at introducing science and English-language education were rebuffed in part because they represented the wishes of an occupying power, and also because their policies had already resulted in indigenous languages and learning institutions being run down.

The fact that both empires gave much support to talented intellectuals who wanted Muslims to engage with modern ideas often ended up having the opposite effect to that which was originally intended. Men such as the 9th-century mathematician al-Khwarizmi, the astronomer Nasir al-Din al-Tusi, or Sayyid Ahmad Khan in India would have been regarded as being too close to unpopular rulers – Khan was knighted by Queen Victoria. The tragedy for many predominantly Muslim societies has been that the voice of science has too often been associated with the blade of a ruler's sword, or the barrel of his gun.

15

Science and Islam: Lessons From History

I have been asking the ulema why their sermons should not exhort Muslims to take up the subjects of science and technology – considering that one-eighth of the holy book speaks of science and technology. Most have replied that they would like to do so, but do not know enough modern science. They only know the science of the age of Avicenna.

Muhammad Abdus Salam, Nobel physics laureate, 1979

It is clear that the colonisation of many developing countries played a part in precipitating the decline of advanced science and learning in the Islamic world. Moreover, the empires of Islam were feeling many pressures from the 16th century onwards, and cost-conscious caliphs would have seen the funding of scientists and scientific programmes as among the first budget items to cut.

Unlike in the modern world, science in the Islamic era was not on the scale that it is today. There were no government departments for science, nor were there science-based multinational companies like Google or Microsoft employing thousands to create the next big thing. Nor were scientists part of enduring institutions, such as colleges or universities. If anything, the genesis of colleges in the Islamic world seems to have been a way to organise those scholars who were opposed to philosophy and rationalism. Knowledge and science in ancient times were supported by individual patrons and when these patrons changed their priorities, or when they died, any institutions that they might have built often died with them. This is a major reason why no observatory lasted more than 30 years in any of the Islamic empires.

Back to the future

Conversations on science's Islamic past often end with a volley of further questions, such as the following: If things were so good before, then why (even in the richer countries of the Middle East) are standards of research, development, discovery and innovation so much lower than those of the developed world? Why are there just two scientists from Muslim countries that have won a Nobel prize in science – Abdus Salam, a physicist from Pakistan who won in 1979, and Ahmed Zewail, a chemist from Egypt, in 1999? Or: Why did science come to an end in the way that it did? If we are talking about countries with large Muslim populations, did the rise and the decline have something to do with religion, or were there other factors? Last but not least: What needs to be done;

what can be done, to revive and give a boost to science and learning?

Overall, the scientific performance today of the 57 member states of the Organisation of Islamic Countries (OIC) is not far off that of some of the poorest countries of the world – this is in spite of the fact that OIC states include some of the world's wealthiest oil-producing nations. Yet if you look at indicators of scientific performance – such as how much is invested in universities; or the quantity and quality of scientific research published in the leading journals; or the numbers of scientists per head of population – then the Islamic world as a whole is not in good shape.

That is not to say that everything is poor, or bad. Certain fields in science and technology are healthy. For example, Iran is a leading nation among the developing countries for its human genetics programme; Malaysia is a leading producer of technology exports; Pakistan is a pioneer in the chemistry of herbal medicines; and Turkey has some of the best universities in the world and its overall scientific output is on a par with that of its neighbours in the south and east of Europe.

Interestingly, the overall scientific situation seems to be a polar opposite of previous times, but there is at least one connection between the present and the golden age of the past – as well as the period of colonial rule. In previous centuries, science benefited hugely from authoritarian leaders. These were men who were not always interested in listening to the public, who used force to silence or eliminate their critics and opponents, and yet at the same time who were keen to push science. They included al-Mamun of Baghdad, as well as the Fatimid ruler al-Hakim. Al-Hakim was a ruler who supported the optics scientist ibn al-Haitham, yet who persecuted a man for refusing to accept the ruler's methodology for compiling an Islamic calendar. Such rulers also included Helagu Khan, who sacked Baghdad yet built the Maragha observatory in Samarkand where the astronomer Nasir al-Din al-Tusi made his contributions to the Copernican revolution. And they also included the many representatives of Europe's nations who would use political and military force to bring modern learning and ideas to the countries that were their colonies.

Today, in those developing countries where science is in better shape, a similar relationship between science and authoritarian rule is very evident. In the countries of the OIC, in Iran, Malaysia, Pakistan and Turkey, for instance, you will find decent salaries for scientists, well-funded research labs, and opportunities for young people to pursue advanced science and learning abroad. The conditions for scientists in non-OIC countries such as China and Singapore are even better. In each case, however, scientists are happy and science is in good shape largely because of strong rulers who are desperate to take their nations into the modern world. At the same time, these are rulers who will not hesitate to use force to override public opinion or undermine their opponents. In the case of the Islamic world, these rulers have included Ayatollah Khomeini, the founder of the modern Iranian state; Mahathir Mohammed, whose strong hand presided over Malaysia for two decades; Kemal Ataturk, the army officer who founded modern Turkey; and a succession of military generals who have ruled Pakistan since the 1950s.

This in no way means that more authoritarian rule is the answer for more science in the developing nations. India, much of Latin America, and of course the nations of the developed world demonstrate that the world's best science happens in countries with a tradition of stable representative government. However, we do need to recognise at least two things. First, that in those countries where strong, unpopular rulers are pushing science today, they are carrying some of the legacy of those rulers who came before them. And second, we need to recognise that in those countries where public awareness of science is low, one reason may be that science is associated with authoritarian rule, and scientists are seen by populations – many of whom will be very poor – as benefiting from autocratic regimes or being close to military rulers.

Did science need Islam?

So science in Islam's empires needed strong yet generous rulers. But did science need Islam, as a faith, in order to progress? And if it did, should we be encouraging more of the peoples of the Islamic world to become better and more observant Muslims, as a way of improving science in OIC countries? This is an argument that is sometimes put forward, particularly by those who believe that the world as a whole is in the grip of moral decay, and that a return to faith will help to make things better. This is also the view of those political leaders who want to see religion and politics in the Islamic world more closely aligned. They argue that, as the golden age of science and learning took place at a time when states were organised and governed under Islamic laws, a return to such ruling systems is what is needed to move science ahead into the future.

The needs and the requirements of Islam clearly did have an impact on the kind of science that was done, and to that extent you could argue that science did at least *benefit* from the coming of Islam. The need for more accurate prayer times, for example, ensured that many more individuals became interested in astronomy, and this led to the creation of the job of time-keeper inside many mosques. Many such time-keepers, moreover, were also keen astronomers, and a few, such as ibn al-Shatir from Damascus, did groundbreaking work. Similarly, a need to help people to calculate inheritance according to Islamic guidance was at least one reason behind the development of algebra by al-Khwarizmi in 9th-century Baghdad. And last but not least, the religion's teachings on healthcare also helped to push the development of medicine and hospitals.

One way of asking whether Islam as a faith was central to the progress of science would be to look at the sources of funding for different scientific institutions, in particular, to see whether scientists and scientific institutions were allowed to be funded by religious endowments. The rulers of Islamic empires created a special endowment fund that was designed to pay for institutions which were important in meeting the obligations of faith. This fund was known in Arabic as a *waqf*, and it still exists in many countries, where it is used to help support the very poorest as well as paying for the upkeep of mosques. If scientific institutions would have had access to these funds, we can safely say that, for all practical purposes, science and faith would have been regarded as one and the same. What historians see from the records of these endowments is that some institutions that had a scientific function were entitled to religious funding, and they included hospitals. Other institutions, however, were clearly prohibited, and they included observatories. Individual scientists were also not allowed to be funded from these sources.

Challenging personalities

A second route to finding out the extent to which the Islamic faith drove science is to look at individual scientists themselves, and to ask whether faith and belief motivated them to experiment, to innovate and to invent and discover new things. What this book shows is that many scientists, engineers and philosophers were indeed Muslim, but also that many were from other faiths. Those that were Muslim, however, were far from being part of the mainstream. The picture that emerges is of a set of individuals who were more likely to challenge received ideas – whether on science or religion – and were therefore not always willing to go with the mainstream. In some cases, the tendency of scientists and thinkers to push at the boundaries attracted charges of heresy. This was at least the case with ibn-Sina and the Andalusian Sufi philosopher ibn-Arabi.

Did Islam need science?

Perhaps one very powerful argument against the view that religion was the major factor in scientific progress is in the nature of Islam itself. Yes, it is true that scientists found all kinds of ways to make the obligations of faith easier to perform, and that this helped to drive scientific discovery. Indeed, we still see faint echoes of that work in the digital compasses you can buy today that point the faithful towards Mecca; or the programmable clocks that can recite the call to prayer. However, at the same time, if you talk to anyone who has converted to Islam, one thing they will tell you is that a key attraction of the faith is not so much its complicated science and technology, but the fact that it carries a simple message, and that its obligations need minimum fuss and expense to carry out. What they like about their new religion is that there is no class of person (neither scientists nor clerics) telling people what they can and cannot do.

So, as we have seen, even today in the age of atomic clocks and GPS navigation systems, Muslims all over the world will still begin the fasting month of Ramadan only after the crescent of the new moon has been spotted with the naked eye. Similarly, in hot countries, the faithful will still use the length of a shadow to work out times for praying; and no one minds if someone kneeling down in the direction of Mecca might be a few degrees out. Not only that, but we know that many of Islam's major mosques from the past are not exactly aligned towards Mecca.

Islam and new knowledge today

This brings us to another – and often controversial – issue: whether Muslims (today or in the past) have had difficulty in accepting new knowledge, especially knowledge that contradicts earlier teachings. And, if this is the case, whether it might be an additional impediment to the development of science and learning. In other words, did religion have anything to do with the decline?

Historians are divided on this issue. Some argue that al-Mamun's inquisition is at the root of many of today's problems. If he hadn't forced rationalism on his peoples, then his critics would not have organised themselves into guilds and colleges, where science was often not part of the syllabus. Others say that many, many key developments in science and innovation took place after this event, and also after the Sufi theologian al-Ghazali's polemic against ibn-Sina. And they argue that Sufism, far from being anti-science, produced one of the most ambitious theories to try to explain the nature of reality – this was ibn-Arabi's theory that God and all of life might be part of a giant inter-related super-organism, which he called 'the unity of existence'.

One important lesson from the past is that Islamic societies were receptive to hearing and discussing new ideas, even if they didn't always agree with them. Until the 15th century, scientists from the Islamic world themselves were generating much of this new thinking. When this process of indigenous learning slowed down and moved to Western Europe, events such as the Copernican revolution were still widely accepted in the Islamic world. Even the publication of Charles Darwin's *Origin of Species* in 1859 was discussed and published in the media of many Islamic countries.

How can science return to the nations of the Islamic world? In many countries, much progress is already being made. But to achieve developed-world standards, governments and those with influence will need to do at least three things. There need to be massive investments, both in educating people and in building institutions. This will be hard for the poorest countries and they will need help, both from their wealthier neighbours and from the broader international community. Second, governments need to give their peoples the freedom to inquire, and the freedom to innovate. And third, science must never be used to attack people's freedom to believe.

The empires of Islam created the conditions for a staggering renaissance in science and technology, some of which undoubtedly helped the scientists of Western Europe. Yet those caliphs and rulers who were most enthusiastic about science were also harsh on their critics, and used science and new knowledge to force people to make choices in religion. If science is to return to the nations of Islam, it must do so without interfering in people's freedom to believe.

Timeline

570–632 **The life of the Prophet Muhammad**

Muhammad and his followers move to Medina in the Hijra.
  Later set as Year 1 of the Muslim calendar (622)
Death of the Prophet Muhammad (632)

632–661 **Islam is led by the four Rightly Guided
  Caliphs**

Abu Bakr becomes first caliph (632–34)
Umar becomes second caliph (634–44)
Expansion to Syria
Expansion to Iraq
Capture of Jerusalem (638)
Introduction of the Hijra calendar
Expansion to Persia
Conquest of Egypt
Uthman becomes third caliph (644–56)
Expansion into the Maghrib

651–700

Compilation of the text of the Qur'an into a book begins 632
  to 634. Completed 634 to 644
Ali becomes fourth caliph (656–61)

Assassination of Caliph Ali (661)

***661–750* Umayyad caliphs rule in Damascus**

Umayyad dynasty established in Damascus

Muawiya I becomes caliph (661–80)

The battle of Kerbala and massacre of Hussain, the Prophet's grandson, and his party. A rift opens between Sunni and Shia Muslims (680)

Caliph Abd al-Malik decrees that only Arabic should be used in official documents (690s)

Introduction of Arabic coinage (693)

Khalid ibn-Yazid advises caliph on science

701–750

Islam comes to Spain (711)

Expansion of Muslims into India (712)

Great Mosque of Damascus completed (715)

Crossing of Muslims into France (718)

Battle of Tours (732)

Umayyad dynasty ends in Baghdad (750)

750–800

***751–1258* Abbasid caliphs rule intermittently in Baghdad**

***756–929* Umayyads rule in Spain**

Foundation of Baghdad (762)

Al-Fazari makes the first astrolabe in Islam (777)

Jabir ibn-Hayyan experiments in chemistry

Harun al-Rashid becomes caliph (786)

Introduction of paper industry in the Arab world (795)

The publication industry established as a sophisticated enterprise

The caliph Harun al-Rashid presents Charlemagne with a clock

The Thousand and One Nights makes an early appearance

801–850

The first public hospital is established in Baghdad (809)

The first House of Wisdom is established in Baghdad

Al-Kindi develops cryptography and introduces Indian numerals

Al-Mamun becomes caliph after deposing his brother in a horrific battle in Baghdad (813)

Ziryab the musician arrives in Cordoba (822)

Caliph al-Mamun develops the House of Wisdom (c. 820)

The translation project gets into gear

Al-Khwarizmi promotes Indian numerals and writes his great book on algebra

Medical doctor Hunayn ibn-Ishaq translates Galen

His son, Ishaq ibn-Hunayn, translates Ptolemy

Shammasiyah Observatory set up near Baghdad (828)

Banu Musa brothers publish their book of mechanical devices (850)

851–900

Al-Jahiz publishes *The Book of Animals*

Al-Qarawiyin university established in Fez (859)

Al-Farghani constructs the nilometer in Egypt (861) and publishes his *Elements of Astronomy*

Al-Farabi writes a pioneering book on music theory

Ibn-Firnas makes the first glider flight (875)

Mosque of ibn-Tulun built in Cairo (878)

Al-Battani publishes *On the Sciences of Stars* (c. 880)

Al-Razi identifies measles and smallpox and develops chemical experimentation

901–1000
909–1171 Fatimids rule in Egypt
945–1055 Buyids rule in Baghdad
Al-Zahrawi in Spain writes a manual on surgery (c. 960)
Al-Azhar university is established in Cairo (988)
Ghaznavid dynasty established in Afghanistan and northern India (977)
Fihirst al-Nadim, the catalogue of books contained in the bookshop of ibn al-Nadim (987)
Al-Biruni publishes *India* and *Determination of the Coordinates of the Cities* (c. 990)
Humanist Al-Masudi lays the foundation of human geography
Philosopher and physician ibn-Sina writes the *Canon of Medicine*, the standard medical text for the next half millennium (c. 1000)
The Ghurids succeed the Ghaznavids in Afghanistan and northern India (1040)

1001–1100
Ibn al-Haitham in Cairo experiments with light, reflection and refraction (c. 1020)
1037–1307 Seljuq empire
Poet Omar Khayyam solves cubic equations (c. 1100)
Statesman and educator Nizam al-Mulk administers Seljuq empire, and creates a network of colleges
Theologian and thinker al-Ghazali publishes *The Incoherence of the Philosophers* and directs the college of Baghdad

Constantine translates Greek and Arabic medical texts into Latin

Muslims travel as far as Vietnam where they establish communities

1101–1200

Al-Idrisi of Sicily publishes a detailed map of the world

Philosopher and psychologist ibn-Bajja establishes psychology as a separate discipline

Adelard of Bath translates Euclid from Arabic and al-Khwarizmi's astronomical tables into Latin

Ibn-Rushd publishes *The Incoherence of the Incoherence* and other philosophical works

Gerard of Cremona translates texts from Arabic into Latin in Toledo

Al-Zarqali works on the astronomical *Tables of Toledo* (c. 1160)

Salah al-Din captures Jerusalem (1187) and unites the Muslim world with Egypt as its centre

Al-Hariri publishes his linguistic masterpiece, *The Assemblies*

Yaqut al-Hamawi publishes his *Geographical Dictionary*

Al-Jazari develops the crankshaft and camshaft and designs the elephant clock (c. 1200)

1201–1300

1206–1406 **Mongol empire**

Fakhr al-Din Razi publishes his great *Encyclopedia of Science*

Biographer Abu-Khallikan establishes philosophy of history as a distinct discipline

Ibn al-Nafis puts forward a new theory on the circulation of the blood (c. 1230)

1232–1492 **Nasrids rule in Granada**

Helagu Khan sacks Baghdad (1258); he becomes a Muslim
and builds an observatory

Abbasid caliphate ends (1258)

Nasir al-Din al-Tusi completes his work *Memoir of the
Science of Astronomy* (1261) at the Maragha observatory
setting forward a comprehensive structure of the universe,
and develops the 'Tusi couple' enabling mathematical
calculations to establish a heliocentric worldview

Ottoman empire founded (1281)

Al-Rammah describes gunpowder rockets (c. 1285)

The rise of the Mamluks in Egypt

Islamic science and learning translated into new languages

1301–1400

**1136–1506 Timurids rule in Central Asia and Middle
East**

Ibn-Khaldun writes on sociology and publishes his
Introduction to History

Ibn-Batuta publishes his *Travels*

1281–1922 Ottoman empire

1401–1500

Ulugh Beg builds observatory in Samarkand

Islamic science and learning take off in the rest of Europe

1501–1600

Mughal dynasty established in India (1526)

Eclipse of Timbuktu as the Great City of Learning (1591)

Ottoman architect Sinan builds the Blue Mosque complex in
Istanbul

1526–1857 Mughal empire

Acknowledgements

This book would not have been possible without the efforts of the many individuals and institutions who gave up time, the benefit of their expertise, or who contributed in other ways.

First of all, sincere thanks go to Jim al-Khalili of the University of Surrey, without whose BBC television series this book would still be a distant idea to be pursued in my years of senior living. Thanks also to my agent Peter Tallack and to Simon Flynn and the staff of Icon Books for making it all happen. John Farndon, my colleague on the project and an eminent science writer and editor, turned my text drafts into stylish prose and also provided valuable research assistance. In addition, thank you to Merryl Wyn Davies for producing the timeline, to Alia Masood, Hassan Masood and Seema Khan for diligent proof-reading and fact-checking, and to Hibah Haider for compiling the bibliography.

There is a clear (and unintentional) Cambridge bias in many of the names to whom I owe a lot. I would like to offer thanks to: Sir Brian Heap, Fraser Watts, Julia

Vitulo-Martin and Denis Alexander for giving me the opportunity to study the history of Islamic medicine as a Templeton Cambridge journalism fellow; and to Peter Jones, librarian of King's College Cambridge and Jamil Ragep of McGill University for introducing me to the works of Nancy Siraisi, the foremost authority on ibn-Sina in Western Europe. Thanks also to the inspirational Fatima Azzam of the Islamic Texts Society for advice on ibn-Arabi; and to Yahya Michot (formerly of Oxford University) for teaching a lapsed scientist the rudiments of Islamic philosophy in five unforgettable lectures.

Several institutions helped in ways for which a simple acknowledgement will not be enough. They include: the International Research Centre for Islamic History, Art and Culture in Istanbul, and in particular its founder Ekmeleddin Ihsanoğlu, for material on the history of science during the Ottoman empire, as well as the role of the observatory in Islam; and the Islamic World Academy of Sciences based in Amman, and in particular executive director Moneef Zou'bi who organised a history of science conference in Kazan, Russia, where I had the opportunity to listen to Columbia University historian George Saliba. Thanks also to Ismail Serageldin and the staff of the Bibliotheca Alexandrina for generous hospitality and insights into ibn al-Haitham.

The British Council, in particular Martin Rose and Stephan Roman of the council's Our Shared Europe team, provided time and space to think about Islam in European history. Thanks also to my friend and colleague Mohamed Hassan of the Academy of Sciences for the Developing World, based in Trieste, who invited me to put together a one-day symposium on science and religion for visiting ministers of science and technology from the Islamic world.

Several individuals have encouraged me over the years to sustain my appetite for discovering science in the present-day Islamic world. They are the late John Maddox, David Dickson and Philip Campbell, my editors at *Nature*, and my two long-time mentors, Zia Sardar and Zafar Abbas Malik of Chicago's East-West University.

Last but not least, my family have gallantly put up with frequent absences during the course of writing and researching this book. Alya, Huda, Hibah and Danyal, this book is dedicated to you.

While the utmost care and dedication have gone into checking the facts in this book, any errors, either of fact or of interpretation, are my own.

Sources

Most dates are sourced from the *Concise Encyclopedia of Islam*, by Cyril Glassé (Stacey, 1989) and the *Book of Islamic Dynasties*, by Luqman Nagy (Ta-Ha, 2008).

Asma Afsaruddin, *The First Muslims* (Oneworld, 2007)

Karen Armstrong, *Muhammad: A Biography of the Prophet* (Phoenix, 1991)

Guy Attewell, *Refiguring Unani Tibb* (Orient Longman, India, 2007)

Zaheer Baber, *The Science of Empire* (State University of New York Press, 1996)

Laleh Bakhtiar, *The Canon of Medicine: Avicenna* (Great Books of the Islamic World, 1999)

Jonathan Barnes, *Aristotle* (Oxford University Press, 2000)

Lectures by Amira Bennison

J.L. Berggren, *Episodes in the Mathematics of Medieval Islam* (Springer, 2003)

Eric Broug, *Islamic Geometric Patterns* (Thames & Hudson, 2008)

Ross Burns, *Damascus: A History* (Routledge, 2005)

Ali Çaksu, *Learning and Education in the Ottoman World* (Research Center for Islamic Art and Culture, Turkey, 1999)

Christopher Catherwood, *A Brief History of the Middle East* (Avalon, 2008)

Peter Coates, *Ibn Arabi and Modern Thought: The History of Taking Metaphysics Seriously* (Anqa Publishing, 2002)

Bernard S. Cohn, *Colonialism and its Forms of Knowledge* (Princeton University Press, 1996)

Michael Cook, *The Koran: A Very Short Introduction* (Oxford University Press, 2000)

Michael Cooperson, *Al Ma'mun* (Oneworld, 2005)

Patricia Crone and Martin Hinds, *God's Caliph: Religious Authority in the First Centuries of Islam* (Cambridge University Press, 2003)

Sayyed Misbah Deen, *Science under Islam* (Lulu, 2007)

Ross E. Dunn, *The Adventures of Ibn Battuta* (California Press, 1992)

John Farndon, *The Great Scientists* (Arcturus, 2005)

Al-Ghazali (trans. Tobias Mayer), *Letter to a Disciple* (The Islamic Texts Society, 2005)

George Gheverghese Joseph, *The Crest of the Peacock: Non-European Roots of Mathematics* (Penguin, 1992)

Thomas Glick, Steven J. Livesey and Faith Wallis, *Medieval Science, Technology and Medicine: an Encyclopedia* (Routledge, 2005)

Dimitri Gutas, *Greek Thought, Arabic Culture* (Routledge, 1998)

Heinz Halm, *The Fatimids and their Traditions of Learning* (I.B. Tauris in association with The Institute of Ismaili Studies, 2001)

Michael Hamilton Morgan, *Lost History* (National Geographic, 2006)

Mushirul Hasan, *A Moral Reckoning* (Oxford University Press, India, 2005)

Ahmad Y. al-Hassan and Donald R. Hill, *Islamic Technology: An illustrated history* (Cambridge University Press, 1992)

Ahmed Y. al-Hassan, *Transfer of Islamic Science to the West* (FSTC, 2007)

Salim T.S. al-Hassani (ed.), *1001 Inventions: Muslim Heritage in Our World* (FSTC, 2007)

Lectures by Salim T.S. al-Hassani

John D. Hoag, *Western Islamic Architecture: A Concise Introduction* (Dover Publications, 1963)

E.J. Holmyard, *Alchemy* (Dober, 1991)

Albert Hourani, *A History of the Arab People* (Faber, 2005)

Ekmeleddin Ihsanoğlu and Feza Günergun, *Science in Islamic Civilisation* (Research Center for Islamic Art and Culture, Turkey, 2000)

Ekmeleddin Ihsanoğlu, *Science, Technology and Learning in the Ottoman Empire* (Ashgate Variorum, 2004)

Halil Inalcik, *Turkey and Europe in History* (Eren Publishing, Turkey, 2006)

Muhammad Iqbal, *The Reconstruction of Religious Thought in Islam* (1938)

Ibn-Qayyim al-Jawziyya (trans. Penelope Johnstone), *Medicine of the Prophet* (Cambridge University Press, 1998)

John Keay, *India* (HarperCollins, 2001)

Hugh Kennedy, *When Baghdad Ruled the World* (Da Capo, 2006)

Hugh Kennedy, *The Great Arab Conquests* (Phoenix, 2008)

Sadiq-ur-Rahman Kidwai, *Gilchrist and the Language of Hindoostan* (Rachna Prakashan, India, 1972)

David Landes, *Revolution in Time* (Viking, 2000)

David C. Lindberg, *Theories of Vision: From Al-Kindi to Kepler* (University of Chicago Press, 1981)

Oliver Leaman, *A Brief Introduction to Islamic Philosophy* (Polity Press, 1999)

G. Le Strange, *Baghdad During the Abbasid Caliphate* (Kessinger Publishing, 1883)

Martin Lings, *Mecca: From Before Genesis Until Now* (ArcheType, 2004)

Hafeez Malik, *Political Profile of Sir Sayyid Ahmed Khan* (Adam Publishers, 2006)

Maria Rosa Menocal, *Ornament of the World* (Back Bay Books/Little, Brown and Co., 2002)

Mustansir Mir, *Iqbal* (I.B. Tauris/Oxford University Press, 2006)

Roy Mottadeh, *The Mantle of the Prophet* (Oneworld, 2008)

Seyyed Hossein Nasr, *Islamic Science* (Tajir Trust, 1976)

Seyyed Hossein Nasr and Muzaffar Iqbal, *Islam, Science, Muslims, and Technology: Conversations with Seyyed Hossein Nasr* (Islamic Book Trust, Kuala Lumpur, 2007)

F.E. Peters, *Mecca: A Literary History of the Muslim Holy Land* (Princeton University Press, 1994)

Attilio Petruccioli and Khalil K. Pirani, *Understanding Islamic Architecture* (Routledge Curzon, 2002)

Peter E. Pormann and Emilie Savage-Smith, *Medieval Islamic Medicine* (The New Edinburgh Islamic Surveys, 2007)

Lectures by Peter Pormann, University of Warwick

Roshdi Rashed (ed.), *Encyclopedia of the History of Arabic Science*, Vol. 1 of 3 (Routledge, 1996)

Sayyid Athar Abbas Rizvi, *A History of Sufism in India*, Vol. 1 of 2 (Munshiran Manoharlal Publishers, India, 1986)

Chase F. Robinson, *Abd al-Malik* (Oneworld, 2005)

Adam Sabra, *Poverty and Charity in Medieval Islam* (Cambridge University Press, 2000)

K.G. Saiyidain, *Iqbal's Educational Philosophy* (Muhammad Ashraf Press, India, 1938)

George Saliba, *Islamic Science and the Making of the European Renaissance* (MIT Press, 2007)

Lectures by George Saliba

Abdulaziz Y. Saqqaf, *The Middle East City: Ancient Traditions Confront a Modern World* (Paragon House Publishers, 1987)

Aydin Sayili, *The Observatory in Islam and its Place in the General History of the Observatory* (The Turkish Historical Society, 1988)

Mark J. Sedgwick, *Sufism: The Essentials* (The American University in Cairo Press, 2000)

Zaid Shakir, *The Heirs of the Prophets* (Starlatch, 2001)

Muhammad Zubayr Siddiqi, *Hadith Literature: Its Origin, Development and Special Features* (The Islamic Texts Society, 1993)

Simon Singh, *Big Bang* (HarperPerennial, 2005)

Nancy G. Siraisi, *Avicenna in Renaissance Italy* (Princeton University Press, 1987)

Hugh Tait (ed.), *Five Thousand Years of Glass* (British Museum, 2005)

Richard Tapper and Keith McLachlan, *Technology, Tradition and Survival: Aspects of Material Culture in the Middle East and Central Asia* (Frank Cass, 2003)

F.W. Thomas and R.L. Turner, 'George Abraham Grierson', Proceedings of the British Academy, Volume 28, 1943

Howard R. Turner, *Science in Medieval Islam* (University of Texas Press, 2002)

Manfred Ullmann, *Islamic Medicine* (Edinburgh University Press, 1978)

Tim Wallace Murphy, *What Islam Did For Us* (Watkins, 2006)

W. Montgomery Watt, *Muslim Intellectual: A Study of Al-Ghazáli* (Edinburgh University Press, 1971)

W. Montgomery Watt, *The Faith and Practice of Al-Ghazáli* (Oneworld, 2000)

Salah Zaimeche, *Baghdad* (FSTC, 2007)

Index